The Spitalfields Project

**Volume 1:
the
archaeology**

ACROSS THE STYX

The Spitalfields Project

Volume 1:
the
archaeology

ACROSS THE STYX

Jez Reeve and Max Adams

CBA Research Report 85
Council for British Archaeology
1993

Published 1993 by the Council for British Archaeology
Bowes Morrell House, 111 Walmgate, York YO1 2UA

British Library Cataloguing in Publication Data
A catalogue for this book is available from the British Library

ISSN 014178 1 9

ISBN 1 872414 07 9

Typeset in-house by Lisa Pickering

Printed by Adpower Advertising, Halifax

Frontispiece West entrance of Christ Church, Spitalfields

Front cover The northern parochial vault exposed for the first time in 140 years — looking north from the parochial vault

Back cover Christ Church, Spitalfields before 1822 (permission of the Museum of London)

Contents

List of illustrations

List of tables

Microfiche

Summary

The excavations at Christ Church, Spitalfields unearthed *c* 1,000 burials from 1729 to 1852. Their interrelationship and significance has been examined in detail in seven chapters.

Chapter 1 recounts the history of the site from 1714, covers the political machinations of church and state, the social implications of taxation and labour, the 'hidden agenda' of intramural burial and the construction of the church and the unreliable builders thereof. A description of the varied purposes and uses of the church down to the present day brings the history of the site into an interesting and living perspective for the present day reader.

In Chapter 2 the aims of the project are laid out, to dig beneath the ceremony and place the historical site in an archaeological context. It also describes the pioneering methodology of this first post-medieval crypt excavation which had to cope with unique difficulties such as the threat of smallpox and lead-poisoning. It forms an essential text for any future investigations of this nature and provides a template for the peculiar personnel and logistical management requirements. It also outlines the method of pro-forma data recording.

The central chapter gives a detailed description, with copious illustrations, of the pattern of depositional and post-depositional processes which led to the filling of the crypt.

In the area of historical research great steps have been made in the understanding of 18th and19th century funerary practice, particularly that of the sexton's role and of the undertaking industry. Chapter 4 examines the contemporary attitude to burial, noting the desire for prestige, safety and the permanence of an intra-mural burial. This is set against the reality of 'body-snatchers', church financial constraints and the predeliction of some sextons for alcohol. Insight has been gained into the working method of four or five sextons who worked in the crypt between 1729 and 1867. It includes an amusing account of the colourful circumstances in which external burial ceremonies were sometimes undertaken and how this led to the development of suburban cemeteries, such as Highgate.

The material culture is discussed in detail in Chapter 5, concentrating on the coffins themselves and the coffin furniture, ie the pressed tin artefacts which adorned the outside of the burial container. A view of their style and manufacture is offered with an interpretation of what might be considered the more profitable lines. Other artefacts which are covered are those found inside the coffins and within the soil dumps.

Chapter 6 concentrates on the textiles. The excellent preservation of textiles, in the varied eco-environments across the site, has been especially exciting for costume historians, particularly in the examination of contemporary underwear and night garments.

In Chapter 7 it has been possible to examine the site formation processes, both human and natural, in a detailed way because of the number of known variables. These can be considered transferable, in many instances to other, more orthodox, archaeological sites.

This volume sets the context for the accompanying anthropological study which details the specific observations from the bodies themselves.

Résumé

Les fouilles entreprises à Christ Church, Spitalfields, ont révélé environ 1000 inhumations datant de 1729 à 1852. Leur corrélation et leur signification ont été examinées en détail en sept chapitres.

Le chapitre 1 raconte l'histoire du site depuis 1714, traite des intrigues politiques de l'Église et de l'État, des conséquences sociales des impôts et du travail, de 'l'ordre du jour secret' des inhumations à l'intérieur de la ville, de la construction de l'église et des entrepreneurs guère fiables qui l'exécutèrent. Une description des fins et usages divers de l'église jusqu'à l'heure actuelle permettent au lecteur moderne de voir l'histoire du site dans un contexte intéressant et vivant.

Le chapitre 2 sert à tracer les objectifs du projet, c'est-à-dire de faire des fouilles en-dessous de la cérémonie et de mettre le site historique dans un contexte archéologique. Il décrit aussi la méthodologie originale de ces premières fouilles dans une crypte post-médiévale qui devaient se confronter à des problèmes uniques comme le risque de petite vérole et de saturnisme. Il constitue un texte essentiel pour toute enquête éventuelle de ce genre et offre un modèle des besoins très spécialisés relativement au personnel et à la gestion logistique. Il trace également les grandes lignes de la méthode d'enregistrement des données dans les règles.

Le chapitre majeur donne une description détaillée, accompagnée de nombreuses illustrations, de la structure des dépositions, caveau par caveau. Il montre comment les caveaux étaient remplis pour commencer et comment on les réorganisait par la suite pour mieux utiliser l'espace dont on disposait. Il y a des discussions des procédés suivis pour les dépositions et après les dépositions qui ont éventuellement rempli la crypte.

On a fait de grands progrès dans le domaine de la recherche historique quant aux pratiques funéraires du 18ème siècle et du 19ème siècle et on les comprend mieux, particulièrement en ce qui concerne le rôle du sacristain et l'industrie des pompes funèbres. Le chapitre 4 examine l'attitude contemporaine par rapport aux enterrements, remarquant que les gens veulent avoir un enterrement à l'intérieur de la ville à cause du prestige, de la sécurité et de la permanence que cela garantit. Tout ceci se situe dans le contexte des 'voleurs de cadavres', des contraintes financières ecclésiastiques et de la prédilection de certains sacristains pour les boissons alcoolisées. Nous avons pu mieux comprendre la méthode de travail de quatre ou cinq sacristains qui ont travaillé dans la crypte entre 1729 et 1867. Le chapitre 4 comprend aussi un exposé amusant des circonstances pittoresques dans lesquelles des cérémonies de sépulture externes étaient parfois entreprises et comment cela a mené au développement de cimetières de banlieue comme celui de Highgate.

Le chapitre 5 discute en détail de la culture matérielle et traite en particulier des cercueils eux-mêmes et de leurs équipements, c'est-à-dire les objets façonnés en fer blanc qui ornaient l'extérieur du cercueil. Le chapitre offre aussi une vue du style et de la fabrication ainsi qu'une interprétation de ce qui pourrait représenter les gammes les plus rentables. Le chapitre 5 traite aussi d'autre objets façonnés, comme ceux qu'on a trouvé à l'intérieur des cercueils et des dépotoirs des sols. Le chapitre 6 traite des textiles en particulier. L'excellent niveau de préservation des textiles dans les divers environnements écologiques du site a été particulièrement intéressant pour les historiens du costume, relativement à l'étude des sous-vêtements et vêtements de nuit contemporains en particulier.

Dans le chapitre 7, il a été possible d'examiner en détail les processus de formation du site, humains ainsi que naturels, grâce au nombre de variables connues. Dans de nombreux cas, on peut les considérer comme transférables à d'autres sites archéologiques plus orthodoxes.

Ce volume situe le contexte de l'étude anthropologique qui l'accompagne, laquelle donne le détail des observations spécifiques des corps eux-mêmes.

Übersicht

Bei den Ausgrabungen der Christuskirche (Christchurch), Spitalfields wurden zirka 1000 Grabstätten zwischen 1729 und 1852 gefunden. Ihre Beziehungen zueinander und deren Bedeutung werden hier in sieben Kapiteln in ihren Einzelheiten geprüft.

Kapitel eins gibt die Vorgeschichte ab 1714 des Ortes wider, erzählt von den politischen Intrigen der Kirche und des Staates, den sozialen Implikationen der Steuern und Arbeit, das 'versteckte Bild' der Mauerbegrabungen und die Kirchenstruktur und deren unzuverläßigen Baumeistern. Eine Beschreibung der verschiedenen Zwecke und Benützungen der Kirche bis zum heutigen Tage rückt die Vorgeschichte des Ortes in eine interessante und lebhafte Perspektive für den heutigen Leser.

Kapitel 2 legt die Ziele des Projektes fest: einen Einbick in die Zermonie zu gewinnen und den historischen Ort in einen archeologischen Zusammenhang zu bringen. Das Kapitel beschreibt eußerdem noch die Weg bahnende Methologie der ersten nach-mittelalterlichen Krypta-Ausgrabungen, die mit einzigartigen Schwierigkeiten, wie die der Gefahr einer Pockenerkrankung und Bleivergiftung fertig werden mußten. Es ist ein wesentlicher Teil für alle zukünftigen Investigationen dieser Art und liefert eine Schablone des Bedarfes an besonderen Personal und logistischem Management. Es ist auch der Grundriß eines Pro-Forma Data-Berichtes.

Mit unzählichen Illustrationen beschreibt das mittlere Kapitel im Detail wie die Gruften, eine nach der anderen in ein Schema angelegt waren. Es illustriert, wie die Gruften anfänglich zugefüllt waren und später, wie sie reorganisiert wurden, um von den vorhandenen Platz besser auszunützen. Es gibt auch Diskurse über Absatz-und Post-Absatzprozesse, welche zur Füllung der Krypta führten.

Große Fortschritte wurden im Bereich der historischen Forschung gemacht, die Begräbnissitten des achzehnten und neunzehnten Jahrhunderts zu verstehen, im besonderem die Rolle, die der Küster und die

Beerdigungsindustrie inne hatte. Kapitel 4 untersucht die zeitgenössische Auffassung zu Beerdigungen und verzeichnet den Wunsch nach Prestige, Sicherheit und die Dauerthaftigkeit eines Mauerbegräbnisses. Das Ganze wurde den 'Leichnam-Räubern', den finanziellen Kircheneinschränkungen und der Vorliebe mancher Küster für Alkohol gegenübergestellt. Man gewann Einblick in die Arbeitsmethoden von vier oder fünf Küstern, die dort in der Krypta zwischen 1729 und 1867 arbeiteten. Das Kapitel enthält auch amüsante Berichte über farbenfrohe Umstände, in denen manchmal externe Beerdigungszeremonien abliefen und wie diese zur Entwicklung der vorortlichen Friedhöfe führte wie zum Beispiel Highgate.

Die materielle Kultur wird in Kapitel 5 in Detail diskutiert, die sich auf die Särge selbst und auf die Sargaustattung konzentriert, wie zum Beispiel die gedruckten Zinnkunstgegenstände, die das Äußere des Sarges schmückten. Ein Profil des Stiles und der Herstellung ist gegeben mit einer Erklärung, welche Produktionslinie wohl gewinnbringender betrachtet werden konnte. Andere behandelte Kunstgegenstände sind die, die im Inneren der Särge und innerhalb der Erdklumpen gefunden wurden. Kapitel 6 konzentriert sich auf Stoffe. Die ausgezeichnete Erhaltung der Stoffe, die sich in der ekologischen, vielseitigen Gegend des gesamten Ortes befinden, begeisterte speziell die Trachtenhistoriker. besonders die Forschung der zeitgenössischen Unterwäsche und Nachtbekleidungen.

Da die Zahl der Variierungen bekannt ist, war es im Kapitel 7 möglich den menschlichen sowie auch den natürlichen Entstehungsprozeß des Ortes im Detail zu prüfen. Man findet, daß man diese Prozesse in vielen Fällen auf andere, mehr orthodoxe archeologische Orte übertragen kann.

Dieses Buch bildet den Rahmen für die anthropologische Begleitstudie, die die bestimmten Beobachtungen von Leichnamen selbst ausführlich prüft.

Preface

The materials are abundant, indeed overwhelming, but they are often as ponderous — sometimes as obscure — as the monuments to which they relate. The works on archaeology, too, exceed those on most other subjects, not only in number and bulk, but also in expense. As in the case of our beautiful cathedrals, light only enters through a very costly medium.

(Henry Godwin, *The English Archaeologist's Handbook*, 1867)

The crypt of Christ Church with All Saints, Spitalfields, East London, is the first post-medieval burial vault to have been comprehensively investigated by archaeological methods. Henry Godwin's comment on the cost of archaeology has achieved historic irony: it was written during the year in which the vaults at Christ Church were closed and sealed after 138 years of continuous burial. Excavations, which took place between 1984 and 1986, cost more than five times the amount spent on building the church.

The value of the Christ Church project will ultimately rest on its uniqueness as a piece of anthropological research. Excavations yielded nearly 1000 individuals, of whom more than 400 have been identified. The retrieval of several generations of a single family has offered anthropologists an extraordinary opportunity, in conjunction with copious documentary material, to reconstruct many aspects of life during London's Industrial Revolution.

It is axiomatic that the anthropology of the people of Spitalfields could not stand without the archaeological context which provides its framework. This context reaches back to the point of an individual's death and forward through the processes of deposition and decay to the present day. Through the study and reconstruction of these processes we achieve not only a vivid picture of depositional behaviour in this burial environment, but also provide the essential confidence in provenance without which no anthropological study could claim validity.

In presenting the archaeological context in this report we have chosen to do just that, and leave a full synthesis of the data for the future. It will take many years of analysis to do justice to the vast quantity of information produced by both halves of the project: the archaeological and palaeo-pathological. To this end we offer a large number of data on fiche, and an archaeohistorical framework within which to view these data.

A history of burial in the church forms chapter 1, and relevant historical material is included in appendix A. We have thought it necessary to provide some methodological insight into the actual process of excavation at Christ Church in chapter 2, partly because many difficulties were encountered which are not common to other archaeological investigations and partly to illuminate some of the press and professional speculation which arose over health and safety issues. Chapter 3 provides the backbone of the report: a reconstruction of the depositional sequence in the vaults between 1729 and 1867. The funeral trade of the 18th and 19th centuries, quite apart from being a bizarre and fascinating study in itself, played a crucial role in the development of Georgian and Victorian attitudes towards death, and is discussed in chapter 4. The following chapters deal with the material culture of Christian death strictly from the archaeological evidence, with the intent that the collection of material catalogued on fiche will provide a reference for future studies of the artefacts of post-medieval Christian death. Finally there is some account of the ways in which archaeologists have used primary data from the excavation to reconstruct events in the crypt. This centres on a discussion of archaeological formation processes which have operated on the deposits at Christ Church, and the ways in which these have been used to infer past events. This report is intended to promote study of the evidence retrieved at Christ Church. It should not be thought of as the last word on the matter. We hope that the decision to sacrifice synthesis for speed of publication will be justified.

Acknowledgements

Investigations at Christ Church were funded initially by grants from the Greater London Council, the Wellcome Trust and the Nuffield Foundation. Post-excavation was funded by a grant from the Historic Buildings and Monuments Commission (England).

Excavations were carried out between October 1984 and April 1986 under the aegis of the Incumbent and Parochial Church Council of Christ Church with All Saints, Spitalfields, administered by the Friends of Christ Church, Spitalfields. Post-excavation analysis was administered by John Hunter, Department of Archaeological Sciences, University of Bradford.

The following professional archaeologists worked for the project: Max Adams, Portia Askew, Mark Bowis, David Bowsher, Julie Carr, Susan Cole, Simon Cottrell, Michael Dawson, Liz Dyson, Jenny Lamprell, Cath Mason, David Parkin, Robert Pearce, Deirdre Power, Jez Reeve and Alison Steele. These archaeologists worked to an extremely high standard under difficult and unpleasant conditions.

Members of the church and the Friends of Christ Church, Spitalfields, are to be thanked for their support, patience, and help during the project, especially the Rev Eddie Stride, Eleanor Murray (Chairman), Paul Gray (Administrator), Eric Elstob (Treasurer), James Lloyd Williams (Accountant), and many others, to whom the authors are most grateful. The architect, A D (Red) Mason, gave much valuable time to many aspects of the project and the authors would particularly like to acknowledge his many contributions.

Much generous advice and help was freely given to the project from many sources. The authors particularly wish to thank the following: Richard Morris, for encouragement, advice and support; Julian Litten, for practical help and sound advice; members of the advisory committee, for their expertise; John Hunter, for patience and much help; Rob Janaway, for ideas and support; Harvey Sheldon; Richard Gem; John Earl; Jonathan Balkind; the Bishopsgate Institute; the staff at Pip the Printers on Bishopsgate; Chris Prescott, and others. The help and support given by Gill Chitty has been much appreciated throughout the project.

Contributors to this volume have played an important part in the success of the project. Stuart Davies conducted the excellent historical research which forms the bulk of chapter 4, and gave advice on its uses and abuses. Brian Leddington and Paul Budd researched aspects of the metalwork which have been included in appendix form. Andrew Jones kindly undertook to evaluate insect remains and give advice on their retrieval. John Price and Joy Ede contributed their specialist skills in conservation and wood identification. Photography was conducted by members of the excavation team, by Phil Crabbe, Nicholas Mackenzie, and Malcolm Lind, to whom many thanks are due. Additional drawings were provided by Jeremy Cross, Cath Mason, and Anna Chudecka.

On behalf of the excavation team the authors would like to thank Dr Susan Young and members of the Health and Safety Executive for overseeing the team's health.

1 History of the site

1 The Commissions for Fifty New Churches

> The spiritual flame burned bright in the Church of England at the opening of the eighteenth century, and nowhere more brightly than in London.
>
> (Port 1986, ix)

The Tory general election victory of 1710, towards the end of the reign of Queen Anne (1702–1714), saw a period of celebration in which Tories sought an opportunity to reaffirm the status of the High Anglican Church in England, and particularly in London. It was not simply a dislike of nonconformism and liberalism which prompted their crusade; it was also a perceived lack of provision for organized worship. In the capital a combination of events and actions including the ravages of time, the Great Fire in 1666, and rapid growth which doubled the London population to half a million within approximately 50 years of the Fire, meant that many churches had either been destroyed or were in a decrepit state, and many people, especially outside the City itself, lived in areas where there were no churches at all.

In February 1711 a House of Commons Committee was formed to examine a petition for the rebuilding of the parish church at Greenwich, using funds left over from coal dues extracted for the rebuilding of St Paul's cathedral (Port 1986, xi). This Committee seems to have been used by the Speaker in the 1710 Parliament, William Bromley, as a means to examine the wider question of 'what Churches are wanted within the cities of London and Westminster, and the Suburbs thereof' (*Commons Journals* XVI, 495). Petitions from the metropolitan parishes applied pressure to the Committee and eventually a scheme for building new churches in the capital was drawn up by Francis Atterbury, High Church Prolocutor of the Lower House, on 9 March 1711. This was then commended by the Queen to Parliament on 23 March, and the Commissions for Fifty New Churches were begun.

In the early 18th century the then estimated population of the 26 metropolitan parishes of London and the 7 hamlets of Stepney was about 513,000. Forty-six Anglican churches, tabernacles and chapels, 61 Dissenting and 14 Quaker meeting houses, together with 13 French congregations, provided all the resources for their organized worship. The Committee calculated that, allowing for 4750 persons per existing place of Anglican worship, an additional 72 churches were needed. On consideration of the number of French Protestants (Huguenots) and Dissenters, this was reduced to a convenient figure of 50.

An Act (Statute 9 Anne c.22) raised coal dues additional to those required for the completion of the Christopher Wren's new St Pauls, and required that the new churches should be built 'of stone and other proper materials . . . with towers or steeples to each of them' (*Commons Journals* XVI, 542, quoted by Port 1986, xiii). They were to be grander than the Wren churches of the English Restoration. Commissioners were appointed to acquire sites for the new churches, and to supervise their construction.

A further Act was required to provide for the continuation of the Commission (10 Anne, c.11). This second Act also dealt with the division of the new parishes, and forbade intramural burials. It was thought that the burial of bodies within the walls of the churches was unseemly and injurious to public health (Lewis 1721). The latter clause proved highly ironic in the contexts both of the 1850 Commission (HMSO 1850), which re-examined the practice of intramural burial, and the eventual excavation of the crypt at Christ Church.

The Commission, first under Queen Anne and then under George I and George II, continued in operation until 1758, when it was finally disbanded. Only twelve of the proposed churches were ever built. The fortunes of the Commission can, however, be closely traced (Port 1986) through their minutes (Appendix A). Few of the High Church Commissioners appointed in 1711 were to be allowed to continue after the death of Anne and the first onslaught of the Whig supremacy. Many of the Commissioners had been Tories: lawyers, city magnates, philanthropists and the like. With the Whiggish character of the new Commission after 1716 many of the lay Tories vanished, as did James Gibb, Hawksmoor's co-surveyor and architect of buildings such as St Martin in the Fields. Hawksmoor himself remained, and was ultimately responsible for six of the twelve Commissioners' churches. Of these, Christ Church is now regarded as the finest, although, as noted below, it was certainly not always so.

Up until the second quarter of the 18th century Spitalfields was a hamlet within the parish of St Dunstan's, Stepney. However, by 1715 the hamlet claimed nearly 20,000 inhabitants (Port 1986, xii). Many of these were Huguenot refugees who had fled France after the revocation of the Edict of Nantes in 1685. On 14 November 1711 (Port 1986, entry no.7) the Commissioners decided that the hamlet of Spitalfields in the parish of Stepney was to be divided into two new parishes, those of Christ Church,

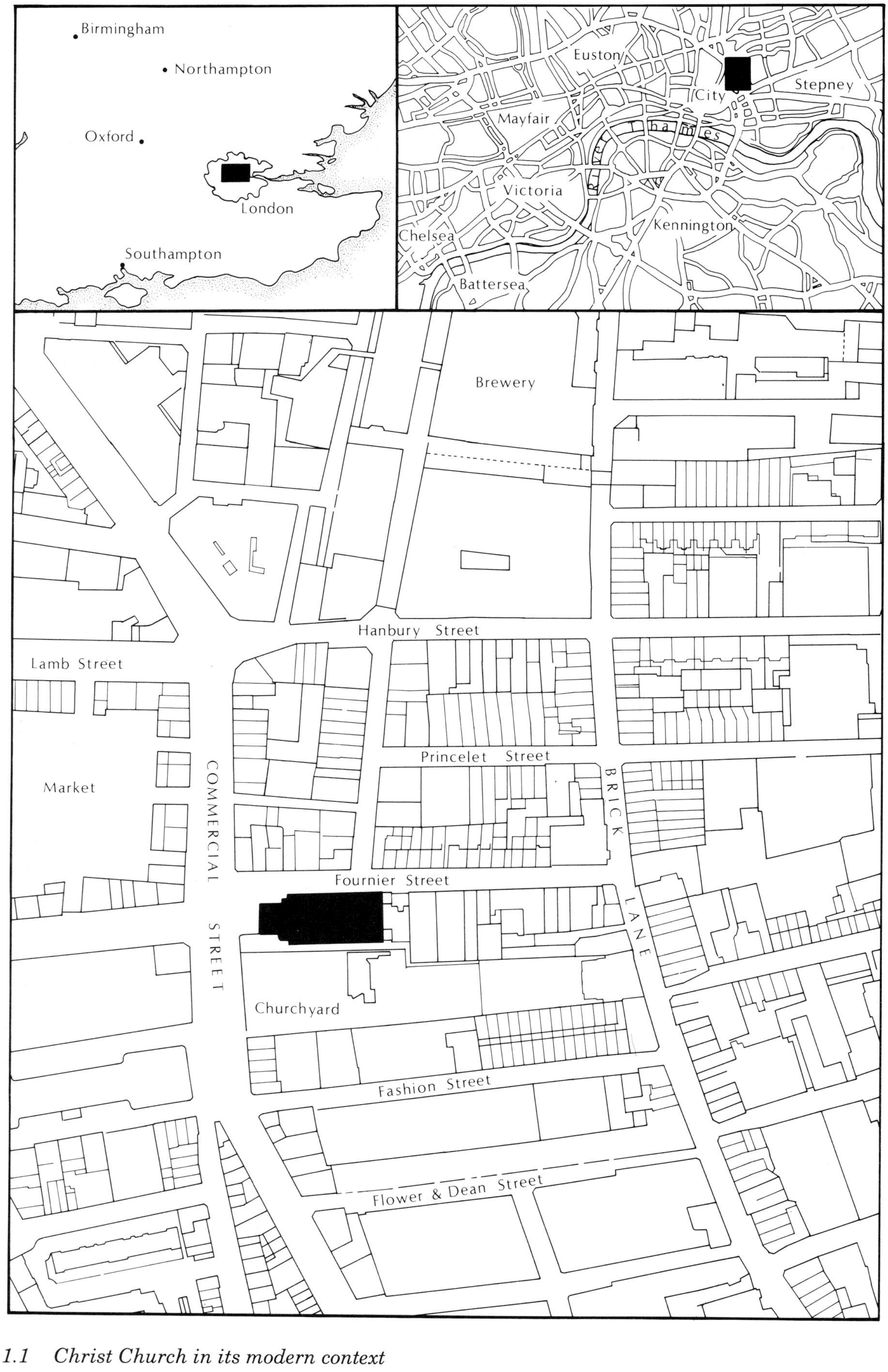

Figure 1.1 Christ Church in its modern context

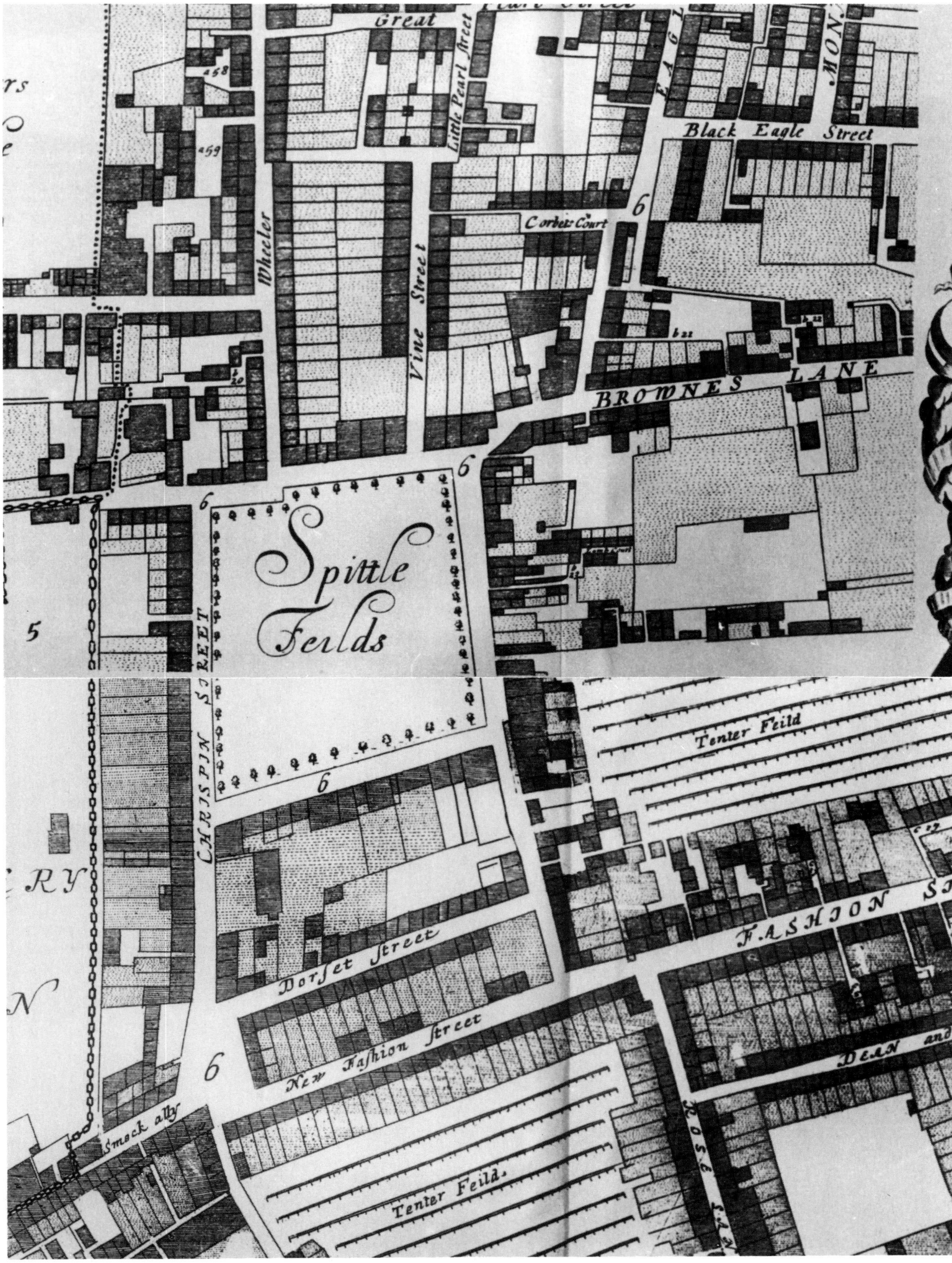

Figure 1.2 Ogilby and Morgan's 1681–2 map showing the hamlet of Spitalfields

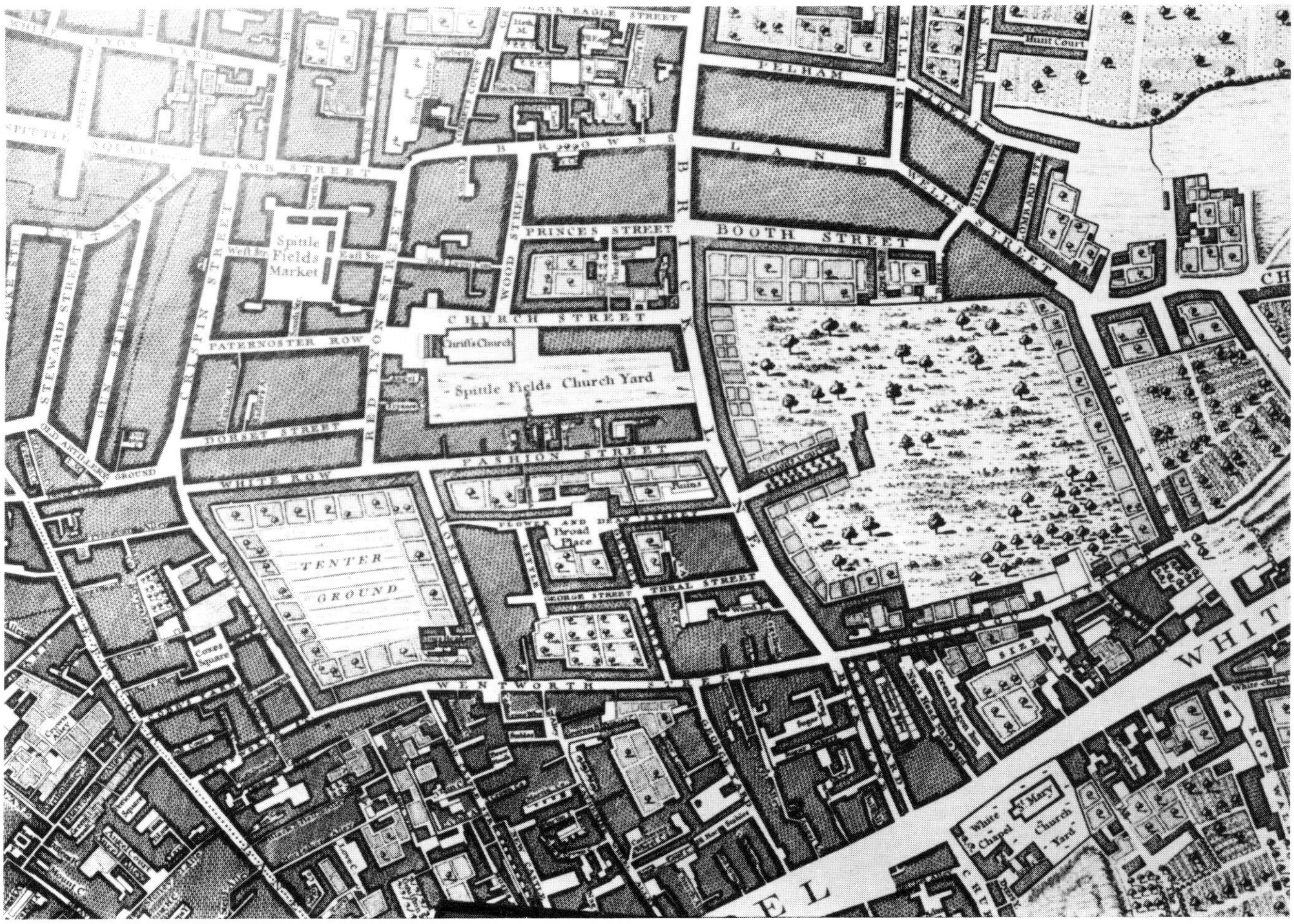

Figure 1.3 Rocque's 1746 map showing Spitalfields and the site of Christ Church

Spitalfields, and St John's, Wapping, respectively, and that two new churches were to be built. Wapping later gained another, Upper, parish at St George in the East. It was, however, not until the spring of 1714, after a representation from several of the inhabitants of Spitalfields, that a design for a church there was drawn up by the Commissioners' surveyor, Nicholas Hawksmoor. On 22 April 1714 the design, estimated to cost £9,129 16s, was approved (Port 1986, 81.1). At the same meeting it was decided that all churches built by the Commission should be raised high above the streets by means of vaulting under the pavements, to reflect, in Vanbrugh's words, the 'awful majesty' of God.

Spitalfields features prominently in the minutes of the Commission. The most important of these references have been reproduced (Port 1986) in Appendix A.

2 The building of the church

Figure 1.1 shows Christ Church, Spitalfields, in its modern context. Ogilby and Morgan's map of 1681–2 (Fig 1.2) shows the hamlet of Spitalfields with the site of Christ Church undeveloped, and the area largely open ground. The site eventually chosen by the surveyors, as shown on the Rocque map of 1746 (Fig 1.3) was in the late 17th century almost entirely open ground, with the exception of one or two 'old, ruinous and ill-tenanted' tenements (Lambeth Palace Library, Deeds, Christ Church, Spitalfields, 24 July 1713). The land had been the property of William Wheler of Datchet, who divided the site between his three daughters. Having passed through several other hands (Sheppard 1957, 152), all three sites were finally purchased for a total of £1260, and the deeds were signed and sealed on 6 November 1713.

A labourer was hired to dig the foundation of the church at a cost of £120, and in June 1714 tenders were invited from bricklayers and stonemasons. The bricklayers Slemaker and Goodchild, who also worked on St Anne's, Limehouse, and St George in the East, were frequently mentioned in the commissioners' minutes, apparently because of the poor quality of their workmen and the bricks which they were using. Despite having been sacked at least once, as the minutes indicate, Goodchild continued to work on Christ Church until 1722, when it seems that his work was completed (Sheppard 1957, 155).

The foundation stone was laid by Mr Edward Peck in 1715 near the south-east corner of the nave. Peck was a dyer of Red Lion Street who was one of the original Commissioners. Sheppard (1957, 155) suggests that he played no small part in arranging the purchase of the site for the church. Indeed, he was honoured by the parish with a large memorial adjacent to the altar, and in April 1727 was granted a family vault (Stepney Central Library Local Collection, Hamlet of Spitalfields Minute, 3 April 1727). This is important in the context of the clause of the first commissioning Act of Queen Anne, noted above, which forbade the use of the crypts of Commissioners' churches for intramural burial.

The work, which proceeded from 1714 continuously for some five years, eventually ran into trouble, as the Commissioners' minutes testify. The coal dues, raised at the rate of 2 shillings and then 3 shillings per chalder on all 'coal and culm' into the Port of London, were evidently insufficient for the ambitious programme of church building. Entry 253, cited below (Appendix A), illustrates the problems facing the surveyors. In April 1720/21 the roof timbers had been lying around through the winter and had suffered to the extent that they had to be tarred before being hoisted into position (Sheppard 1957, 156). This lack of resources, and the consequent debt, were further exacerbated by vandalism. A plumber's bill for 1721/2 included a charge for a labourer hired for 'watching the lead' on 63 nights (Sheppard 1957, 156). In the *King's Collection of Topographical Drawings*, held in the British Museum, more than 30 drawings of various constructions at Christ Church show the extent to which changes of taste and resource availability led to modifications of the design of the building. As late as the mid-1720s there was uncertainty as to whether the finished church would have a steeple.

After 1721 new funds were made available to the extent that most of the labourers who had effectively been subsidizing the building were paid what they had been owed and the surveyors, Hawksmoor and James, had their salaries raised from £150 per annum back to the original £200 which they had been paid between 1712 and 1719/20. Even then, interruptions to the work reoccurred, so that it was not until 5 July 1729, fifteen years after the laying of the foundation stone, that the church was consecrated by the bishop of London, Edmund Gibson. The total cost was £39,162 17s 6d, more than four times the original estimate.

The resultant place of worship was not such a triumph as all the effort, money, and time spent on it would presuppose. By the beginning of the 1730s, well into the era of the Whig Supremacy, High Church pretensions as to the scale and design of the Commissioners' churches appeared unfashionable, even a little grotesque. In 1734 Ralph (1734, 6) remarked that Christ Church had become 'one of the most absurd piles in Europe' due to the 'monstrous expense' lavished upon it. Christ Church was already *passé*.

3 Christ Church, 1729–1984

The vestry minutes for the period 1729–43 do not, apparently, survive. From the 1740s onwards, however, there is ample evidence that the church underwent a series of repairs, renovations, and alterations which have culminated in the programme of the 1970s and 1980s to restore the building to something like its original state. Recorded changes include the removal of the stairs from the north entrance to the church, and the subsequent alteration of windows. The spire was altered in 1822–3, removing the crockets to create a less elaborate form (Colvin 1978, 404). In around 1826 John Leschallas, vestryman, charged £967 15s 9d for painting the church throughout. He seems to be the same John Leschallas who purchased a vault in the crypt for his family.

The period 1836–66 saw the most radical changes to the fabric and structure of the building. On 17 February 1836 a serious fire in the steeple led to the destruction of most of the woodwork in the upper part of the spire, much of the masonry, and the clock. This incident may have been responsible for a deposit of soot and ash which was recorded in some areas of the crypt, and which has implications for the dating of several phases of deposition (see chapter 3). On 3 January 1841 the steeple was struck by a bolt of lightning which caused further damage. There was a thorough internal restoration in 1851, followed by drastic alterations in 1866. These alterations were made by the architect Ewan Christian. They played a significant role in the history of the crypt, and especially its phase of abandonment. Much of the material introduced to seal the vaults appears to have been derived from debris produced during this activity (see chapter 3). (There is a comprehensive discussion of many aspects of the history of the building in volume XII of the Survey of London: Sheppard 1957.) Other local activities may have been significant. The creation of Commercial Street between 1843 and 1857 may have generated refuse deposits, some of which could have been introduced into the vaults (see chapter 3).

Following the decision during the 1970s to restore the building to its former glory the interior has been stripped, much of the plaster moulding reconstructed, and the masonry of the portico renovated. The graveyard has been cleared, and presently accommodates a school, adventure playground, and ornamental garden. No archaeological record of the graveyard and its memorials is thought to exist.

4 The crypt

Figures 1.4–1.7 show, respectively, architectural plans of the church and crypt, an axonometric reconstruction of the crypt from the south-west, and a split level plan of the excavated area.

The crypt at Christ Church extends beneath the entire area of the building, including the steps leading up to the portico at the west front. Like the

Figure 1.4 *Ground plan of Christ Church*

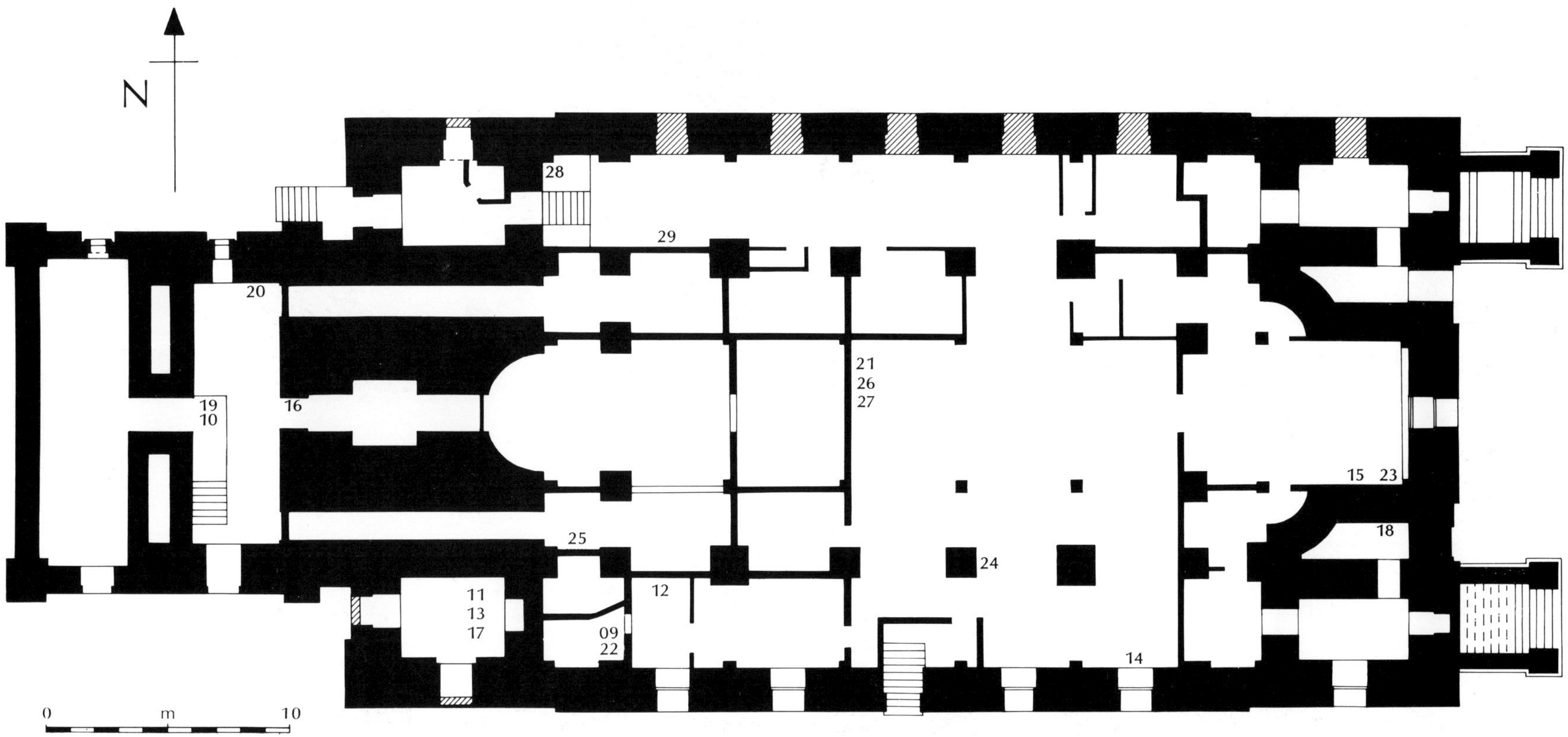

Figure 1.5 Ground plan of the crypt at Christ Church

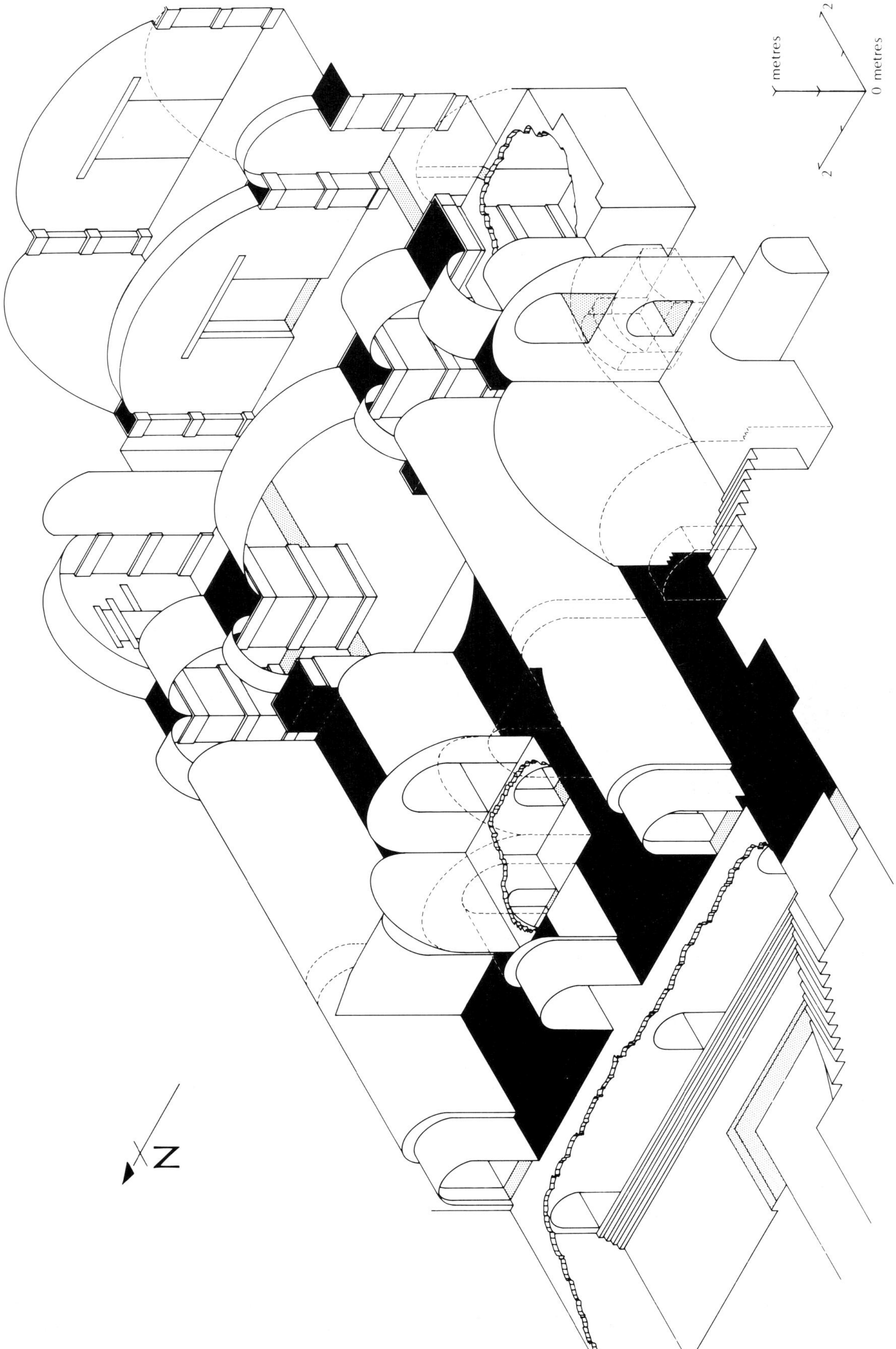

Figure 1.6 Axonometric reconstruction of the crypt from the south-west

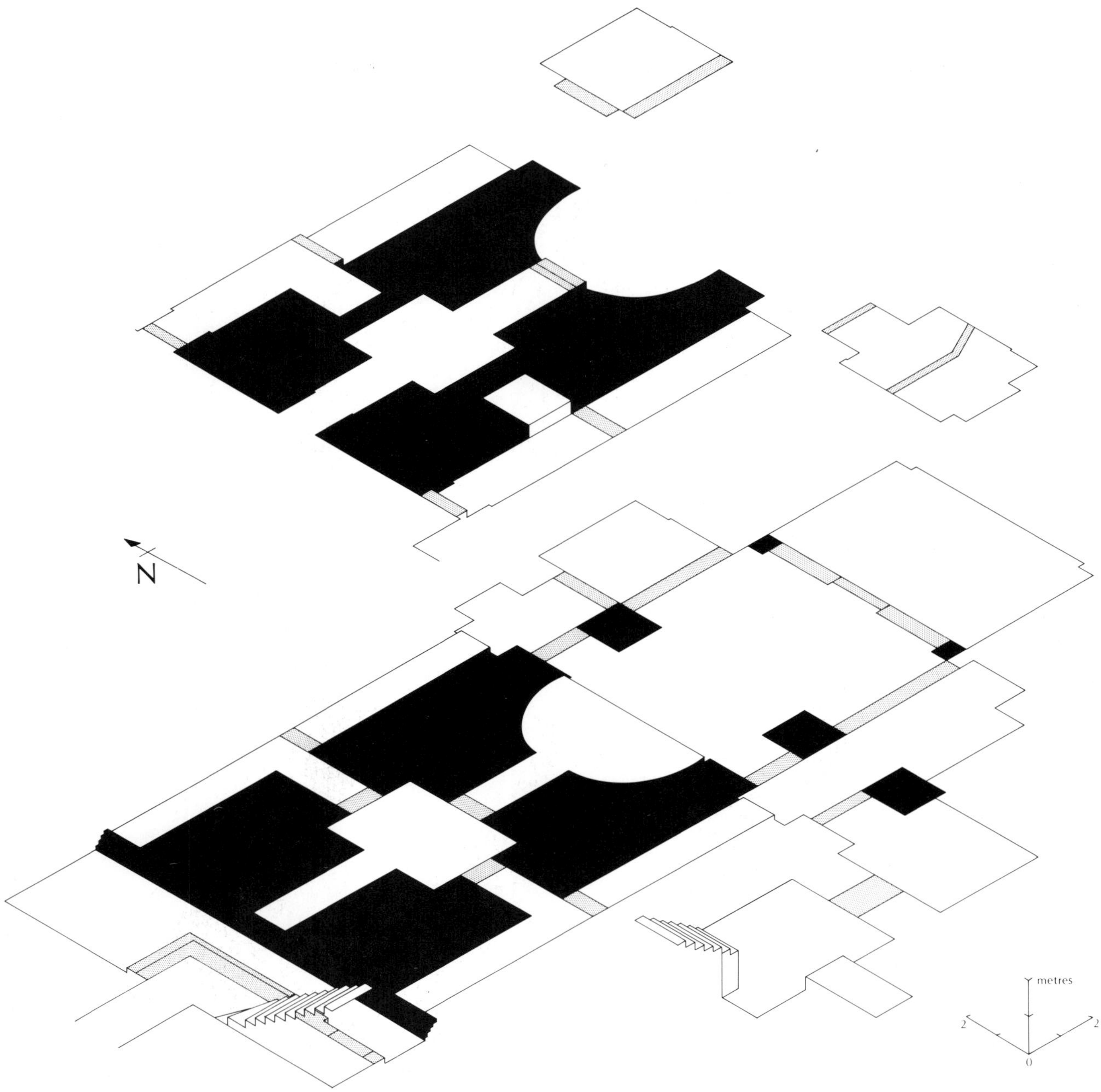

Figure 1.7 Split-level plan of the excavated areas

upper part of the building, the crypt was constructed in two distinct phases. The tower foundation was erected before, and separate from, the main body of the building. This is evident where the apse on the east side of the tower foundation is butted against the vaulting of the main parochial area (Fig 1.8). The tower, weighing 19,000 tonnes, is supported by a brick pier which is pierced by two tiers of tunnels, north, south, and central. The tunnels are themselves vaulted and are apparently part of the original design of the structure. They are of the 'railway tunnel' form described by Litten (1985). The west ends of the three lower tunnels appear to have been secured with doors during the construction phase, and across the entrance to the apse on the east side there are remains of beam slots in the walls which are inferred to have supported a screen or shutter. There are signs of closure on the east ends of the lower tunnels, which appear to have held door frames, and these must have extended upwards to enclose the upper eastern ends of the tunnels. The Commissioners noted in 1720 (see Appendix A, 253.7) that much damage was being done to the works by the ingress of 'the mob', and it seems likely that some provision was made for the storage of equipment at the site within the enclosed area of the tower foundation.

The seven bays of the crypt correspond to those of the main body of the church. There are north and south aisles and a central section, more than 5 m across, which corresponds to the nave. At the east end the form of the chancel is repeated. The supporting

Figure 1.8
Apse at west end of parochial vault showing where the body of the church abuts the tower

piers and external walls are of Portland stone, the floors, internal walls, and vaulting are of the problematic bricks supplied by Slemaker and Goodchild. One exception is the floor of the Hebert vault, which is flagged with Portland stone. The possible reasons for this are discussed in chapter 3. This vault was constructed after the original building phase. The Simpson and Hebert vaults were horizontally divided by a brick vault rising from springers on two piers. An inscription (Appendix C25) gives a *terminus ante quem* of 1843 for the first use of the upper of the two vaults, but it is likely that the alteration occurred considerably before this date.

The first burial which is known to have taken place in the crypt occurred on 3 August 1729, less than a month after the consecration of the church. However, the intent to bury can be dated earlier. The endowment of a vault for Edward Peck and his family indicates that burial had been planned certainly before 1727. The first burial to take place in the graveyard at Christ Church was entered in the parish burial register for 8 July 1729. A further 68,000 or so entries were made in the register before the final entry on 23 February 1859. It is by no means certain that all of the entries represent burials in the vaults or the churchyard at Christ Church, but a figure of several tens of thousands of burials is not uncommon for post-medieval churches in London.

Nine hundred and eighty-three burials were retrieved from the crypt during the excavations. The minimum number of individuals interred was somewhat larger. Two, or perhaps three, private vaults still in existence at the east end of the church remain intact, and will not be cleared in the foreseeable future. These may contain up to 20 further individuals. Parts of individuals, especially skulls, which were retrieved from deposits and could not be assigned burial numbers, accounted for perhaps another 60 possible individuals. If these represent all of the interments in the crypt, then the excavated remains represent approximately 1042:1062, or 98.12% of the total. However, there is a possibility, which should be considered, that other interments were cleared from the crypt prior to the excavation.

Anecdotal information abounds which refers to the removal of decomposed bodies from the eastern half of the crypt at various stages during the last 100 years. Local tradition suggests that the east end may have been used as a rifle range during the First World War, although there is no archaeological evidence for this. Workmen who were involved in the clearance and renovation of the eastern half during the early 1960s give contradictory accounts of the removal of human remains. These clearances took place before the Faculty Jurisdiction Measure of 1964, which was introduced to control proposals for work conducted in churches, although a faculty would still have been required under the Burial Act of 1857; this has not been found. No Home Office Licence for the removal of such remains is known to exist either.

That interments were deposited in the eastern half of the crypt there is little doubt. Inscribed stones bear witness (Appendix C) to individual interments. A slab on the east-facing side of the east wall of the eastern parochial vault states that it was closed, being full, in 1845. Since burials in the crypt are known to have taken place into the 1850s, it is likely that the area to the east of this slab was used as a third public or parochial vault.

Adjacent to this slab is another, cited below in chapter 2, which indicates that upon closure in 1867 the crypt was 'cleared and improved' as nearly as possible. It also states that all unenclosed coffins should be covered with soil and then enclosed by stonework or brickwork. The excavated area was entirely sealed by such means. This implies that all those coffins remaining in the eastern half of the church were placed within a discrete area and then sealed, with the result that the eastern half of the crypt could still be used for other purposes. There is some archaeological support for this hypothesis. The Leschallas vault on the north side of the parochial area was sealed by a horizontal deposit of rubble and concrete after 1852 (Adams & Reeve 1987, 249). Subsequently the upper part of the west wall of this vault was dismantled and entrance obtained through the north wall of the north chasm (NC). Some 100 interments, all of which appear to have been secondary depositions, were then introduced above the Leschallas interments, and into the north chasm, from the north entrance. Since the two parochial vaults had by 1852 already been closed, these coffins probably came from either a third parochial vault, or from a series of private vaults. A combination of the two seems likely. Dates on these secondary deposits range from 1752 to 1847, and it is therefore unlikely that they all came from a parochial vault which would have been used from 1845 onwards. The authors infer that unenclosed family groupings, together with public interments after 1845, make up the 100 or so interments placed in the 'north chasm' and 'north parochial' vaults after 1852. The 'south chasm' also contains redeposited coffins, which may be of this period, as does the lower portico vault. The conclusion drawn is that after 1867 there were no unenclosed interments in the crypt and that the excavation retrieved all those interments which remained at the closure of the crypt in 1867.

None of the vaults at Christ Church was purpose-built. All of them were created by the addition of partitions after the main construction phase was completed. There seem to have been two entrances from the churchyard to the crypt. One of these was the entrance used by the excavators (see Fig 1.9). This entrance was certainly still in use after 1849 when an interment was deposited in the lower portico vault, but seems not to have been used after the closure of the burial areas in 1867. Thereafter, and probably since the 1730s, an entrance in the centre of the south wall of the church appears to have been used. The parochial vaults, and all those vaults to the east of the tower foundation, appear to have

Figure 1.9 *South-west exterior of the church, showing the entrance used by archaeologists*

Figure 1.10 Interior of the crypt beneath the steps leading to the portico at the west end, looking south

been filled from this entrance, rather than from that of the south-east corner. The implications of this are detailed in chapter 4.

Little is known of the uses made of the eastern half of the crypt after 1867, when Queen's Council prohibited intramural burials in the church (see chapter 4). In 1965 the area was converted for use as a rehabilitation centre for 'alcoholic vagrants'. The accessible part of the western half, which included the area below the portico steps (Fig 1.10) and the upper level to the west of the tower foundation, has been variously used as a shelter during the Blitz of 1940, and for the storage of tools. A scene taken by photographer Bill Brandt in 1940 shows a vagrant sleeping in a 'stone sarcophagus'. This was later broken up and used as hardcore at the entrance to the eastern half at the north-west corner of the building. This 'sarcophagus' may be that which contained the remains of Edward Peck who died in 1736. The south-western entrance to the crypt was reused as the entrance to the excavations after 1984. The western half of the crypt has now been cleared of all interments and currently contains material from the church being stored pending the completion of the restoration, after which part of the area will be incorporated into the rehabilitation centre, and the rest used to house services such as heating and lighting for the restored church.

2 Account of the excavation

> Archaeological research has established that from early times, if not from the beginning, steps were normally taken to dispose of the the body with appropriate ritual.
>
> (Poulson & Marshall 1975, 15)

> How far it is justifiable to rifle the tomb for purposes of archaeological interest may be an open question, but even the most hardened delver must have had some doubts on the subject . . .
>
> (Puckle 1926)

1 Setting up the project

In the 1970s a restoration programme was planned for Christ Church which included a scheme for installing heating and other services under the floor in order to avoid compromising the architectural integrity of the church fabric (see chapter 1). The eastern half of the crypt had already been cleared and was being used as a refuge for 'alcoholic vagrants'. The western half, including two large parochial vaults, was still occupied by an unknown number of burials. An initial 'search' was made of the vaults to ascertain their contents by drilling several 'keyholes' through the brick linings and taking still photographs and video film. A proposal for the excavation and scientific examination of the contents of the undisturbed part of the crypt was drawn up in May 1983, emphasizing the extraordinary potential for archaeological and anthropological research offered by the site (see section 2 below). Commercial clearance of the vaults was considered by the Incumbent and Parochial Church Council, but they generously agreed to an archaeological excavation, provided that it could be funded by grants. A Faculty was obtained on 31 October 1983 and a Home Office Licence was granted under the Burial Act 1857, Section 25 (No 15210).The following working practices were stipulated in the Licence:

1. The removal shall be effected with due care and attention to decency.
2. If necessary, freshly made ground lime shall be freely sprinkled over the coffins, the soil and elsewhere.
3. The removal shall be:
   (a) to the satisfaction of the Chief Assistant Environmental Health Officer for the London Borough of Tower Hamlets, and
   (b) in accordance with such additional requirements as may be imposed by the Health and Safety Executive.
4. The remains shall be forthwith conveyed to a laboratory to be constructed within the confines of the church for the purpose of scientific examination.
5. On completion of such examination the remains shall be placed in suitable containers which meet the requirements of the said Chief Assistant, Environmental Health, and shall, without undue delay, be cremated in accordance with the Cremation Acts 1902 and 1952 and the Regulations made thereunder, and in any intervening period, they shall be kept safely, privately and decently.

Excavation commenced in November 1984. Funding was obtained from the Nuffield Foundation, the Wellcome Trust, and the Greater London Council (GLC) Historic Buildings Panel. When an extension to the excavation schedule was required in May 1985 and again in October 1985 funding was received from the Nuffield Foundation and the Historic Buildings and Monuments Commission (HBMC). The post-excavation work and preparation for publication costs have been covered by the HBMC. The project was administered by the Friends of Christ Church, Spitalfields, during the period of excavation, and transferred to the administration of the School of Archaeological Sciences in the University of Bradford after the post-excavation work had begun.

2 Academic objectives

A comprehensive archaeological excavation of a site of this period and type had never previously been attempted. It was hoped that it would yield a high quality and quantity of archaeological and anthropological data, since the crypt had remained substantially undisturbed since its closure in 1867. However, research priorities were limited by the availability of resources and by the data collection policy adopted to optimize these resources.

The objectives can be divided into three main areas: methodology, human biology and post-medieval archaeology. It was anticipated that the excavation of a mortuary site which was recent in archaeological terms would contribute to general archaeological knowledge, particularly in terms of excavation procedure, concepts of context identification and more especially the interpretation of dated material within an archaeological context. It was hoped to assess the validity of some common archaeological preconceptions about the interpretation of stratification from mortuary sites, and determine whether a new approach to the understanding of artefact movement within a mortuary site would be necessary. A reconstruction of the sequence in which burial deposits were introduced and transformed in the crypt might offer some indication of how mortuary deposits are formed (see chapter 7). One of the major objectives in approaching the site archaeologically was to secure a reliable sample of aged and sexed human material. It was expected that coffin plates of various types — breastplates, inscriptions, sideplates, etc — would survive, bearing biographical information which physical anthropologists would be able to use to check the present criteria employed in the assessment of age and sex in human bones.

Archaeological objectives included the collection of data for the study of the undertaking industry of the 18th and 19th centuries, in particular any variations in the burial facility and funeral furnishings throughout the 138 years during which the crypt was used for burial. Although there have been some studies of the material culture of Christian death (eg Morley 1971; Rahtz 1981; Litten 1985), they have been based on either scant or unrelated material. Excavation provided an opportunity to study the management of the crypt by sextons and their assistants. The large quantity of funeral clothing and textile fittings which was anticipated represented a potentially rich collection for specialists studying costume of this period. It was thought that there might be archaeological confirmation of variability in the treatment of the dead during the period of burial here. Local historical research was planned to assess the validity of inferences made from the archaeological data, and vice-versa, thus making it possible to place the material culture from Christ Church into its historical context (see chapter 4).

3 Health and safety considerations

> When the living body is exposed to . . .
> putrid emanations in a highly
> concentrated state, the effects are
> immediate and deadly; when more diluted
> they still taint the system, inducing a
> morbid condition which renders it more
> prone to disease in general . . .
>
> (Lewis 1850, 9)

This is a disturbing view of the risks involved in examination of post-medieval burial vaults, which two years of excavation at Christ Church only partly dispelled. A year was spent in consultation with the Medical Officer of Environmental Health for Tower Hamlets, the Public Health Services Laboratory, the Health and Safety Executive (HSE), the Advisory Committee on Dangerous Pathogens, and the Smallpox Eradication Unit of the World Health Organisation (WHO) in Geneva, to generate an appropriate health and safety code before the excavation commenced. The major cause for concern at this point was the possibility that spores of viable smallpox virus might survive in human tissue (Razzell 1976). Eventually a procedure based on the creation of 'clean' and 'dirty' working areas, was compiled. This involved the wearing of protective clothing and following of safety procedures in the 'dirty' area, with clear guidelines as to where this extended. The procedure was adjusted to cope with unforeseen hazards during later stages of the excavation. All members of staff were subject to medical surveillance throughout the period of excavation.

Although the nature of the work proposed was similar to that done by those professional undertakers who specialize in the wholesale clearance of vaults, it was thought that scientific examination of bodies might present unusual hazards. Undertakers engaged in wholesale clearances of vaults do not follow such severe health and safety practices as those introduced for the excavations at Christ Church.

The health and safety code specified that protective clothing should be worn by all members of staff on site: overalls, hats (including hard hats when necessary), steel-toe-capped wellington boots, and surgical gloves. It also stated that there should be two sorts of protective respiratory masks: anti-dust and micro-biological. These should be worn at all times on site, or when required. When staff had to leave the working, 'dirty', environment they first had to enter the adjacent 'clean' area, remove the protective clothing they were wearing, and wash their hands and face, before then exiting to the street. Shower facilities were made available in case of gross contamination and heavily soiled protective clothing was burnt regularly with the excavated coffin wood at a site approved by the HSE. First aid facilities were provided in all the areas of work and a record kept of all accidents, near-accidents, and illnesses. Members of staff appointed their own health and safety representative.

Medical surveillance was carried out by an Environmental Health Officer who gave an initial explanation of the health risks and necessary precautions which had to be taken. This included advice on tetanus immunization and the threat of anthrax: workers who were allergic to penicillin were advised to have an anti-anthrax inoculation. A doctor from the HSE took samples of blood from all excavation personnel to act as control samples to be used for comparison should any problems of contamination or illness occur. This initial visit was followed by intermittent checks, which were stepped up when the health risks were considered to have worsened. The possibility of lead poisoning was not initially envisaged, and the more insidious health risks caused by the crypt environment were not, initially, fully realized: little air movement, high dust levels, poor visibility, restricted working areas, and the morbid nature of the archaeological material meant that maintaining morale and health was at times difficult. Excavators experienced lingering colds, 'flu',

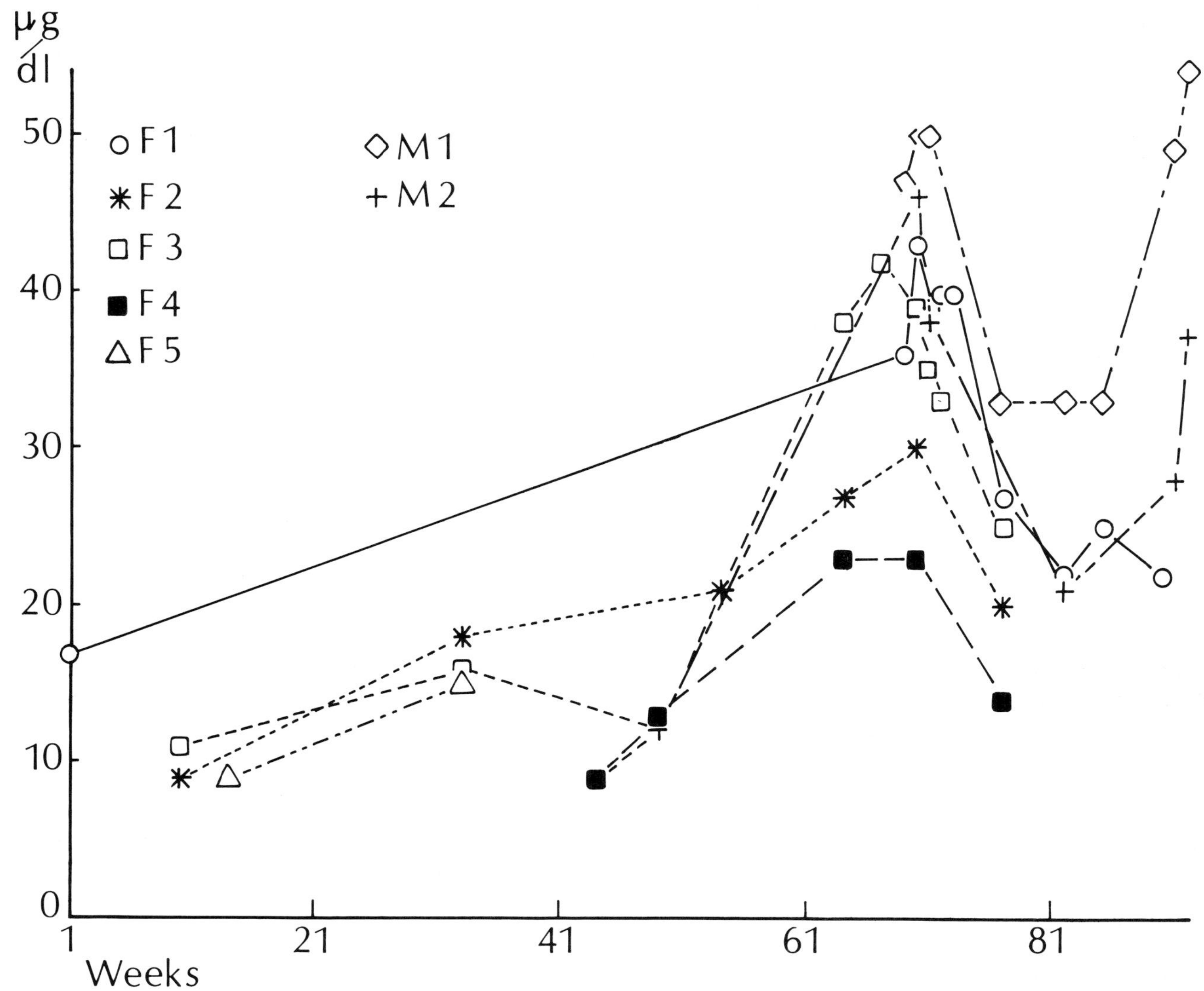

Figure 2.1 Chart showing lead\blood levels in staff during excavations

depression, lethargy, and niggling minor infections throughout the period of work. Dr Waller Lewis, visiting more recent interments in 1850, commented that he had experienced discomfort from the 'putrid emanations' coming from the rotting corpses:

> My investigations have not yet been extended so far as I hoped would have been the case; but the very foul state of the atmosphere contained in the greater number of the vaults has had so depressing an effect upon my health that it became quite impossible to venture on a repetition of inhaling the poisoned air till some of the effects of the previous visit had disappeared.
>
> (Lewis 1850, 29)

In the planning stages no especial provision for sick leave, other than the average one and a half days per month, was made. The unusual circumstances of the work meant that this provision was inadequate. Of the 12 long-term members of staff, 6 terminated their contracts before time and 5 others worked only short periods of between one and eleven weeks. The nature of the archaeological material also meant that in cases of personal bereavement long absences from site were necessary. A major consideration in choosing staff was the need to form a team which could work well together, both professionally and socially, in the unpleasant environment: staff often had to deal with putrescent or mummified human material. Throughout the excavation the greater part of the excavation team was made up of female staff, who within a non-hierachical working structure could create the supportive framework needed to cope with the fluctuations of health and morale.

The main cause for concern in the original health and safety discussions was the danger of smallpox survival. It was decided that the possibility of this occurring was remote, but that precautions should still be taken to reduce any possible risk. When smallpox pustules were found on a semi-desiccated body in April 1985 the continuation of work was forbidden by the HSE until such time as a sample, taken by their officials, was tested and found to be unviable. It took six and a half weeks for this to be demonstrated and prompted speculation in the medical press (eg Zuckerman 1984; Baxby 1985; Dixon 1985; Hopkins 1985; Jessamine 1985; Last 1985; Thomas 1985). Despite the negative results of this examination the HSE decided that all staff and future applicants had to be inoculated against smallpox by a doctor from the Communicable Diseases Surveillance Centre (CDSC). This could only be carried out on individuals with primary inoculation scars, in order to lower the risks of possible side effects. Applicants who had not been inoculated before, or who lived with children or pregnant women, or who were pregnant themselves, were automatically prohibited from working on the excavation.

It became apparent towards the end of the first part of the excavation (in June 1985), when work had started on the investigation of the sealed lead coffins, that lead/blood levels had risen in staff. A month after the beginning of the second phase of excavation (in November 1985) a rise in lead/blood levels was detected in new members of staff. A representative from the HSE provided the team with information about legal maximum lead/blood levels, which are lower for women of childbearing age (to protect the potential unborn foetus) than for men, at 40 ug/dl and 80 ug/dl respectively. Lead can be absorbed 'after inhalation of dust or fume, and after ingestion. Inorganic lead compounds are not absorbed through intact skin, but if they reach the hands they may be ingested' (Cooper 1974,126). The symptoms of mild lead poisoning are 'characterized by lassitude, constipation, abdominal discomfort and pain, anorexia, occasionally nausea, sleep disturbances, irritability, pallor and anaemia' (*ibid*). Some of these symptoms were experienced during the excavation, but they were not attributed to mild lead poisoning at the time.

When sealed lead coffins were opened on a daily basis, lead/blood levels rose more rapidly. Precautions were taken to avoid contamination, through the use of recommended protective clothing. However, as the lead/blood levels in staff always increased when they were working on site it was concluded that the concentration of lead in the atmosphere and in the unexcavated dumps exceeded that for which the recommended safety wear could provide adequate protection. In April 1986 some female members of the staff had to be withdrawn from the working area because their lead/blood levels were over the legal safe limit (see Fig 2.1). When enquiries were made of the HSE representatives as to what ill effects men might experience with high lead/blood levels it was ascertained that high lead/blood levels might be injurious to their reproductive capacities. As a result all members of staff were removed from site when their levels reached the maximum allowance for women. Tests were needed to ascertain what level of lead was present in the atmosphere and unexcavated dumps to gauge the appropriate level of respiratory protection required. These tests were never carried out. A restriction which had been imposed after the smallpox scare, that anyone entering the work place should have a recent smallpox immunization, deterred the relevant specialists from taking the necessary precautions so that they could enter the working environment to take samples. They were not prepared to let their equipment be used by members of staff already inoculated. The only method of dealing with the problem was to continue to use the protective wear provided and to remove staff from site when they reached the maximum level of 40 ug/dl, and keep them off site until their lead/blood levels dropped.

Some thought went into the problems presented by the necessity of lifting and moving heavy objects, notably lead coffins. Many lead coffins weighed more than a quarter of a tonne and often had to be moved some distance before excavation could commence. Mechanical lifting apparatus was not available, so it was done manually. Care was taken to avoid back injury and the dangers of dropping heavy lead coffins. There was always an initial discussion as to whether levers or rollers could be successfully employed to facilitate the move, then the route of transport would be agreed, after which the webbing slings, by which the coffin was to be lifted and carried, were manoeuvred into position. Most importantly, all those cooperating in the manoeuvre would have to agree to act as a team and be prepared to alert other members of the team or be alerted by them to cease operations if at any point unforeseen difficulties were experienced. Four workers were usually used to lift or transport a lead coffin, although sometimes six were required when the route involved a passageway or sharp incline. Scaffolding planks were used as levers and sawn-up lengths of scaffolding poles as rollers to facilitate the moving of heavy coffins along low narrow tunnels. On one occasion a block and tackle was used to lift a lead coffin up a flight of stairs. This was successful, but the method was inappropriate in other circumstances.

Of the minor illnesses whose source could be attributed to the working conditions, ear and throat infections, sinusitis, persistant stomach trouble, back injury, and conjunctivitis were predominant. In one instance a member of staff contracted a skin disease which was never identified, despite considerable medical effort. It was suggested that it could have been similar to a disease noted on tree surgeons, and therefore may have been contracted from fungal and insect remains found in decaying coffin wood.

It is clear that disruptions and physical discomfort or illnesses which occurred during the working period affected attendance (see Fig 2.2), general morale, and possibly at times the efficiency of recording and retrieval procedures. Some of the problems could have been ameliorated, if not avoided, by detailed research prior to excavation, particularly into the effects of deprivation from natural light and fresh moving air in cramped working conditions,

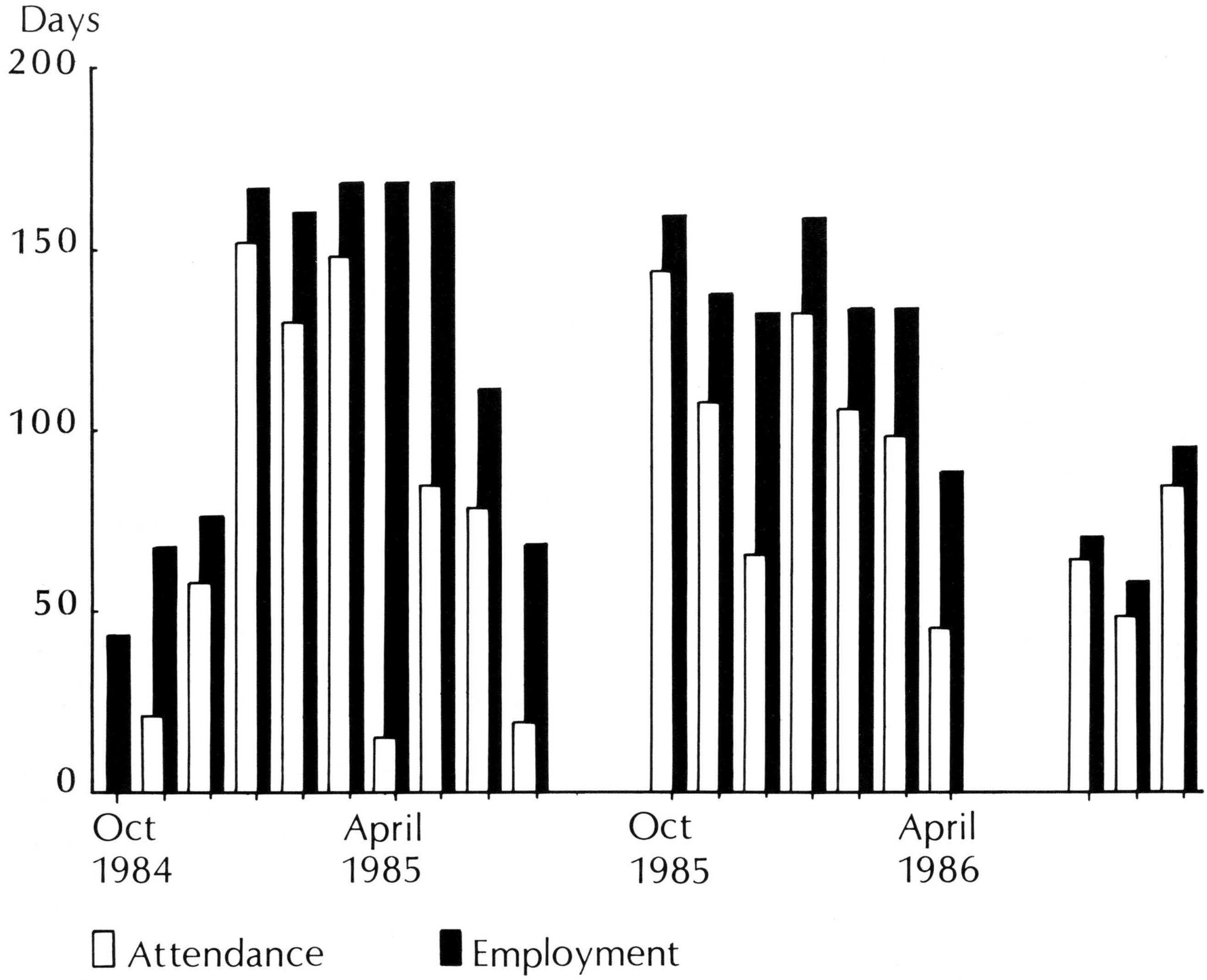

Figure 2.2 Chart showing attendance figures for excavations

such as those encountered by mineworkers and noted in Sick Buildings Syndrome.

4 Methodological approaches

There were no precedents for recording methods appropriate to a site of this type and therefore a procedure specific to Christ Church was devised. Although there are directions on the treatment of graves, grave cuts, and skeletons from archaeological contexts in other methodologies, the archaeological environment at Christ Church was so significantly different from other burial sites that they were not instructive. At Christ Church, in the absence of graves cut into a stratigraphic sequence, coffined burials themselves formed the stratification, onto which dumps of secondary refuse had been deposited. It was important in defining the status of a coffin to unite the concept of a grave with the physical concept of a layer, and with the role of the coffin as artefact. The analogy of a pit was employed in approaching the space represented by the walls of the crypt. The procedure which was devised was based on the single-context system used by the Department of Urban Archaeology (DUA) at the Museum of London (MoL 1980), although considerably redrafted. Planning conventions had to be rethought and specific directions for recording coffins invented. It was decided that from the start an integrated procedure was necessary, to be used by both physical anthropologists and archaeologists. This was drawn up in the form of a report before work began and was used as a manual in conjunction with the several pro-forma context recording sheets, signing-in books and charts and photographs to record the excavation data (Reeve 1984).

The pro-forma recording sheets were designed to facilitate the fastest possible data recovery, concentrating on a minimum set of recording requirements, but providing the possibility for recording of further detail in specific circumstances. There are four headings under which the data retrieval from burial context must be considered:

1 Provenance and physical nature
2 Descriptive information — facility
3 Descriptive information — artefacts
4 Descriptive information — human body

Under each of these headings there is a list of criteria by which every burial context had to be considered. To ensure that the criteria also included the research priorities later set out in the post-excavation research design (Reeve & Adams 1986), pro-forma recording sheets were designed with both compulsory prompts, covering the minimum recording requirements such as location and description, and optional recording prompts, covering the need for extended description required in specific circumstances. Each component related to a signing-in book in which a tabular checklist for each plan, context, artefact, and skeleton was created. Space for recording salient correlations and dated signatures was filled in on the completion of each part of the recording sequence. A hard-backed signing-in book of the 'Analysis' type was used. In this way the location of any plan, context sheet, artefact, or body could be traced on its passage from the field to the laboratory. The books were used as quick reference aids during the excavation and particularly in post-excavation for retrieval of salient information before the generation of a computer database was completed. A strict order of activities was outlined in the manual with directions on when the columns in the books should be signed. The columns in the signing-in books were as follows:

The plan book

Plan no/Context no/Initials & date/Context no/ Initials & date/Context no/Initials & date/No of sheets/ Checked, initials & date

The context book

Context no/Initials & date/Plan no/In workshop, initials & date/Finds/Samples/Other drawings/ Photo/ Finished in workshop, initials & date/ Checked, initials & date/Removed, initials & date

The finds book

Context no/Comment/CF?/ Obj(ect)/ Mat(erial)/ Initials & date/ Photo/ Checked, initials & date/ Removed, initials and date/ Where to/ Analysis/ Returned, initials & date

The skeleton book

Context no/Comments/Initials & date/In workshop, initials & date/Outer coffin no/Inner lead no/Inner wood no/Adult or child/Sample/ Drawn/Photo/Out of vault, initials & date

The site was divided into 5 m grid squares and contexts were referenced with the appropriate south-west corner coordinates. If the context continued into more than one grid square the fact was recorded on

Figure 2.3 View looking west down south-east tunnel (SE:0000.01)

both the plan sheet and in the plan book. Although single-context recording normally dictates one context only to be drawn on each planning sheet, more than one coffin was planned on each sheet, provided that they did not have any stratigraphic relationship one with another. The over- and underlying stratigraphic relationships were recorded on the sheet for each coffin. In order to reconstruct the three dimensions of each coffin, the uppermost side was planned and levels taken at the same time of contiguous coffins in the same row or layer, and after excavation the same data were recorded for the lowermost side. This procedure was changed midway through the excavation to speed the procedure: the uppermost side of each coffin only was planned and the height of each coffin was recorded on the context sheet. Certain problems were encountered in planning some coffins because of difficulties of access. Often the coffins could not be reached from above because they were crammed into low tunnels (Fig 2.3), or stacked up to the ceiling (see Fig 3.12). One way of recording the furthest point in these cases was by using a carpenter's rule with a plumb bob attached to 'feel' out the extent of the coffin. The exact location of the coffin was thus established, and it could be planned once it had been removed and measured. In the main parochial vault, when pressure on time was at its greatest, the area was gridded out with string at 2 m intervals and a draughtsperson then planned all the visible coffins in by eye from a high vantage point. This was independently checked for accuracy by a second member of the team.

A set of pro-forma context recording sheets was designed to record the different types of burial contexts which were anticipated: a sheet for outer coffins, a sheet for inner lead coffins, a sheet for inner wooden coffins or inner wooden lids, a sheet for stratified human remains, and a sheet for all individuals excavated.

The first three sheets were used in conjuction with the context book for recording the coffined interments. Although context sheets using a multiple choice format were designed prior to excavation, these were redesigned almost as soon as excavation began in order to incorporate a prompting system. The context numbers for burial contexts, deposits, interfaces, and structural contexts were allocated context numbers 1–1000 and 4000+. Each of these recording sheets consisted of a folded piece of high quality A3 paper which presented four A4 pages for pro-forma recording prompts. Pages 1, 2, and 4 were identical in each case. Page 1 offered space for the recording of the provenance and type of the coffin context (see Fig 2.4). A diagramatic representation of a coffin was provided so that dimensions and distinguishing characteristics could be recorded quickly, such as the pattern of the upholstery pins, how many kerfs were employed, and the construction in general. The position of the body in the coffin was recorded on page 2 if the diagrammatic representation of the coffin on page 1 had been filled in with distinguishing characteristics.

For the recording of an outer wooden coffin, whether it was an external shell or an entire coffin, the whole of page 3 was used to provide pro-forma prompts (see Fig 2.5). The space for 'Interpretation' was filled in during the excavation of the context. For the recording of an inner lead coffin the upper half of page 3 was used (see Fig 2.5). When a taxonomy for the different forms of construction was created the type of construction was recorded in a two-figure code (0–99). As the pro-forma prompts for the recording of an inner wooden coffin were on the same page, in the case of an inner lead coffin, one or other could be deleted. For the recording of an inner wooden coffin, the same formula as that used for the outer wooden coffin was displayed and completed.

Page 4 presented a printed sheet of metric graph paper on which scaled drawings and sketches could be made of important details, relationships, or associated artefacts.

For the recording of a skeleton two sheets could be used in conjunction with the skeleton book. Every skeleton had to be recorded on the Skeleton Vault Data (SVD) Sheet and only those which had no container were recorded as stratified contexts using the MOL skeleton sheet. Each skeleton was allocated a context number (2001–2999 and 5000+) prior to excavation; only this and biographical information, reduced to a numerical code, were recorded on the SVD sheet by the excavator. The SVD sheet was then used by the physical anthropologists to record the samples which were taken, together with any photographs, date of X-ray, and computerization. In this way there could be no accidental duplication of context numbers and no bias introduced in the use of the methods for ageing and sexing.

Each excavator was responsible for the collection and initial cataloguing of artefacts and samples. Each artefact could be identified by the context number from which it came and a four-letter description code: eg a breastplate was *brpl*, an escutcheon *escu*. A waterproof label was marked in permanent ink with these details, the site code, and the material type, and was included in every case with the packing of the artefact or sample. When taxonomies for the designs of the coffin furniture were created midway through the excavation (see below), the relevant design number was also included on the label. The excavator was also responsible for recording the finds and samples in the finds book. Although each item did not receive a discrete number it was individually labelled and recorded at source, minimizing any discrepancy between the field and finds departments.

As mentioned above, the procedure for excavation and recording was formulated before the excavation had commenced. It was initially anticipated that a large number of the coffins would be sufficiently intact to allow removal to a 'workshop' area, after having been planned, and there be investigated and recorded, without disrupting excavation on site. This situation did not occur very often; as a result most of the excavation and recording of coffins was done on site. This change in excavation policy did

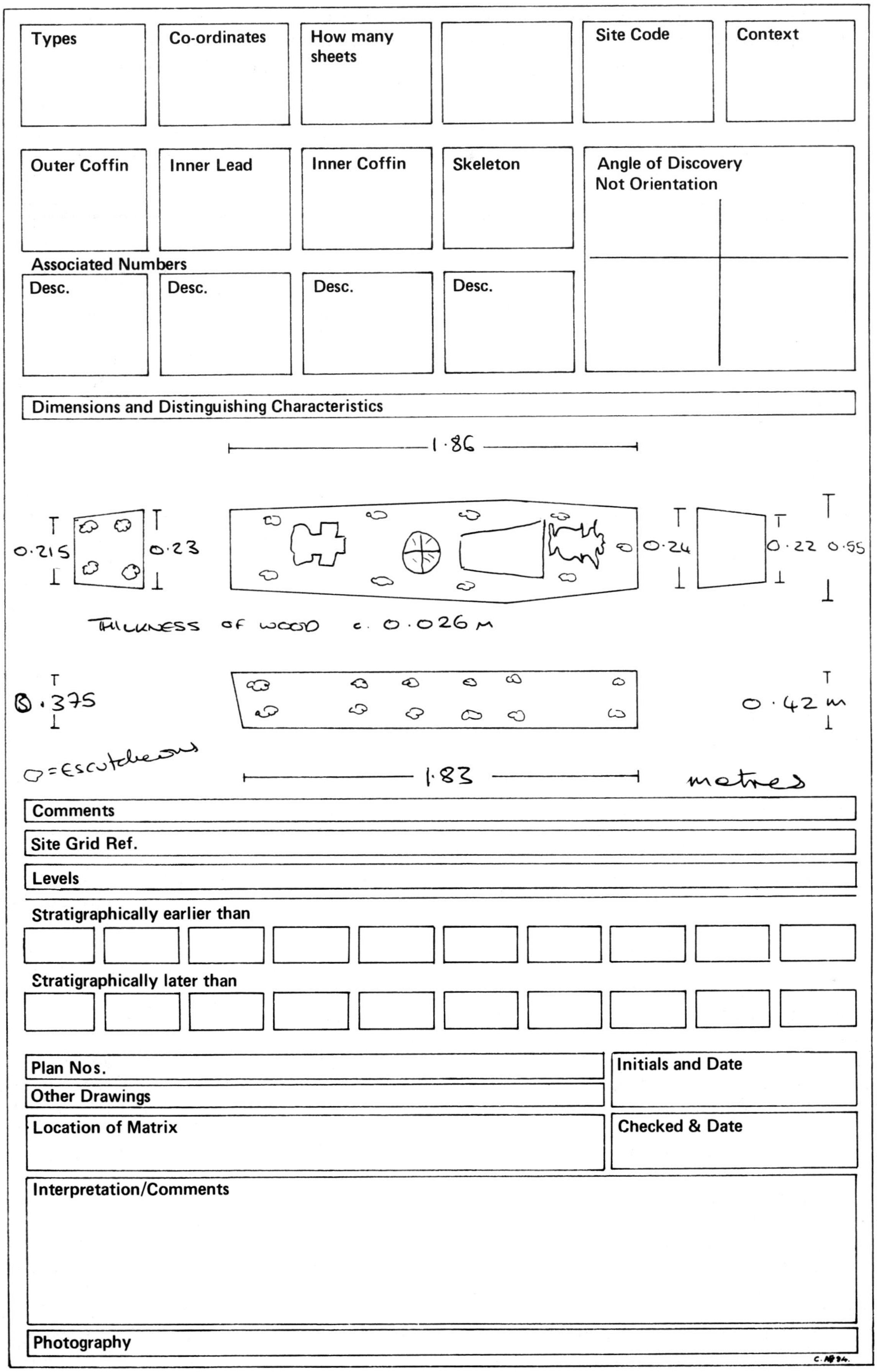

Types	Co-ordinates	How many sheets		Site Code	Context

Outer Coffin	Inner Lead	Inner Coffin	Skeleton	Angle of Discovery Not Orientation

Associated Numbers

Desc.	Desc.	Desc.	Desc.

Dimensions and Distinguishing Characteristics

Comments

Site Grid Ref.

Levels

Stratigraphically earlier than

Stratigraphically later than

Plan Nos.

Other Drawings

Location of Matrix

Initials and Date

Checked & Date

Interpretation/Comments

Photography

Figure 2.4 *Context recording sheet, page 1*

OUTER COFFIN | **Sample Ret.** ☐ | **Context**

Recording Date | Site Code

Associated Nos.

Condition

Breast Plate ☐ | **Legible** ☐ | Inscription

Fab. | Lead ☐ | Ormolou ☐ | Brass ☐ | Tin Plated Pewter ☐ | Gilt Tin ☐ | Iron ☐
Silvered Tin ☐ | Enamelled ☐ | Other | **Sample Ret.** ☐

Shape of Breast Plate | Square ☐ | Rectangle ☐ | Shield ☐ | Tapered ☐ | Lozenge ☐
Cartouche ☐ | Oval ☐ | Other

Covering ☐ | **Sample Ret.** ☐ | Colour | Serge ☐ | Velvet ☐ | Other

Construction | Butt Joint ☐ | Bevelled Edge ☐ | Pitched ☐ | Other

Grip Plate ☐ | **Fab.** | Ormolou ☐ | Brass ☐ | Tin Plated Pewter ☐ | Gilt Tin ☐ | Iron ☐
Silvered Tin ☐ | Enamelled ☐ | Other | **Sample Ret.** ☐

Grip ☐ | **Fab.** | Ormolou ☐ | Brass ☐ | Iron ☐ | Other | **Sample Ret.** ☐

Design | Swag of Roses ☐ | Horse Shoe ☐ | Cherubs Head ☐ | Plain ☐ | Other

Grip Bolts ☐ | **Corner Bolts** ☐ | **Type** | Gilt Brass ☐ | Brass ☐ | Iron ☐ | **Sample Ret.** ☐
Other

Lid Motifs ☐ | Desc. | **Sample Ret.** ☐

Escutcheons ☐ | Desc. | **Sample Ret.** ☐

Upholstery Nails ☐ | Brass ☐ | Gilt Brass ☐ | Silvered Brass ☐ | Iron ☐ | **Sample Ret.** ☐
Enamelled ☐ | Other

Fixings ☐ | Panel Nails ☐ | Screws ☐ | Hinges ☐ | Lace ☐ | Other
Sample Ret. ☐

Timber Type | **Sample Ret.** ☐

Interpretation/Comments

Figure 2.5 *Context recording sheet, page 3*

not affect the procedure for the recording of coffins. It was decided that any attempt to obtain a photographic record of every context, or even of every burial, was impossible for two reasons. First, the speed at which the excavation had to be conducted in order to retrieve 983 individuals at a rate of up to 30 per day in the later stages meant that the time spent on setting up even the most rudimentary photograph for each one would effectively render one member of staff unavailable for burial retrieval. Second, given the difficult conditions for lighting and access which existed in the crypt and the highly variable quality of the data, it was thought that a more selective policy was required. Unfortunately no guidelines were available at the start of the excavation for the photography of burial vaults, and the recommendations made by Phillips (1976) rely heavily on financial resources which were not available to the excavators. Nevertheless the authors are acutely aware that the resulting photographic record is barely adequate for a site of this nature. It is clear in retrospect that photographic recording was given too low a priority and that inadequate provision of materials limited the scope of the photographic policy.

There was no budget allowance for photographic equipment. The excavators used their own cameras: a Pentax SP 1000 35 mm SLR with various lenses, and a Praktica MTL5 35 mm SLR. A medium format camera was sought but could not be obtained. Phil Crabbe of the British Museum (Natural History) undertook to photograph some subjects where the 35mm format could not provide sufficient detail. The artefactual record was taken by Malcolm Lind using a Hasselblad 500 medium format camera with 135 mm macro bellows lens.

The conditions inside the crypt were such that photographs produced very poor contrast, especially between skeletal material and decayed coffin wood. About 90% of the photographic archive consists of colour transparencies. These were on daylight 400 ASA Ektachrome film, unfiltered. Many of these have since been colour balanced, producing an adequate, if not spectacular, result. There are approximately 700 transparencies. Of the monochrome photographs taken, some five rolls were stolen in January 1986 and not recovered.

For working purposes the vaults were lit with 150 watt festoon bulbs which could be moved between areas. Photographs were taken, where possible, using two photoflood lamps of either 275 or 500 watts each. These proved adequate for most purposes. A record card was designed for the on-site photography, which contained details of film, speed, lens, lighting, date, subject, and photographer. Each shot was then given a unique code, as follows: the site code, followed by a two-letter subject code, followed by a context number, plus the sequential number of slides of that code, eg CAS 84:PV:1345.01. The area codes are given in Appendix H. Two complete copies of the on-site photographic archive exist. The paper record has since been integrated with the computer database.

5 Progress of the excavation

In August 1983 the time estimated for the completion of the excavation of the entire crypt was seven months. One month was allowed to set up operations and six to excavate the crypt, with a staff of one supervisor and six excavators. This proved to be a serious underestimation. Discussion of the health and safety procedure held up the start of the excavation until October 1984 (see section 3 above). It was known that the interments had been covered in 'soil' even before the initial 'keyhole' photographs were taken, from an inscription on the eastern face of the wall enclosing the parochial vault (Appendix C27). However, it was not at that point realized how much redeposited material would have to be excavated or the problems such a large deposit would pose to the logistics of finishing the excavation within six months. After a month of excavation (December 1984), when only four members of staff had been secured, it was realized that some way of speeding up the procedure had to be found. The on-site coffin context sheets were redesigned to make recording quicker and more efficient and there were discussions between the archaeological team and the anthropological team about the level of on-site recording for human remains. This resulted in the advice that the excavators had only to record the position of the bones on discovery and their state of preservation (from A–C) before bringing them to the physical anthropological workshop for further examination (see chapter 7). It was also decided that the anthropologists should be responsible for taking any organic samples from the human remains. These alterations to the procedure and the hiring of a further four excavators accelerated the excavation, but by March 1985 only a third of the total estimated number of interments had been excavated, despite the fact that over half the total number of vaults had been cleared.

As described above, the discovery of possible smallpox pustules in April 1985 arrested excavation for one and a half months, during which time some background research was done and design taxonomies created for lead coffins, funereal metalwork, and upholstery pin designs. Resources were found for three more months' excavation, but only another two vaults were cleared as the number of staff dwindled and more labour-intensive lead-shelled coffins had to be excavated.

Initially it was decided that any lead-shelled coffins were not to be opened, but only externally recorded, taken away and buried intact. However, as most of the lead shells were in a good state of preservation, with attached legible coffin plates bearing biographical information, they represented a source of material of a high inference quality (IQ), especially for the physical anthropologists. Thus all the lead-shelled coffins were excavated after further negotiations with the Health and Safety Executive. The second phase of excavation started in October 1985 after a gap of two months in which another six months' excavation grant was secured and six staff were appointed. More than half of the remaining interments were encased in lead and approximately three-quarters of the entire tonnage of spoil had to be removed from five of the remaining vaults. In addition to these difficulties lead/blood levels began to increase (see section 3 above) and staff had to be suspended from on-site work. Under these circumstances it was deemed necessary to speed up the excavation procedure still further. Taxonomic sequences were used when possible for the recording of the upholstery pin design, the lead coffin construction, and all the funereal metalwork, and only the coffin plate bearing biographical information was collected. By the end of January 1986 the level of recording was reduced further still so that all component details for multiple-shelled coffins were recorded on the one outer coffin sheet, mostly in numerical code form. Despite these drastic reductions in information retrieval, designed to maximize the sample of human material with biographical details for the physical anthropologists, some specialist sampling still continued. Where textiles survived in a good state of preservation entire costumes were retained, and when insects and parasites were found these were kept for analysis.

Even after the completion of excavation the Health and Safety Executive insisted that the entire crypt be cleared of all extraneous material and washed in a virucide solution before the risk of contamination could be eliminated and the human material be released for study. This was carried out by the excavators. The excavations produced 983 individuals, nearly 4000 small finds, more than 100,000 items of numeric data, 16 tonnes of lead, and 250 tonnes of rubble.

6 Post-excavation

Post-excavation commenced in earnest in October 1986. Two staff in the Hanbury Street office coordinated with specialists and the anthropological team at the British Museum (Natural History). The primary purpose of the team was to present the Christ Church data as quickly as possible to the archaeological world in the light of the increasing number of post-medieval burial vaults which were and are being cleared without archaeological examination. It was hoped that by creating a precedent and a collection of material, Christ Church might pave the way for a greater archaeological input into what is a very neglected field of study.

It was thought that an essential component of the production of data was the creation of a relational computer database. Data of this type lend themselves particularly well to a digital recording system. Very large numbers of variables which are quite impossible to manipulate manually are readily digested by a database. It is unfortunate that computer facilities were not available before the excavation. Nevertheless, an IBM PC 'AT' machine was eventually obtained by kind permission of the Friends of Christ Church for the use of the project. Initially no software was forthcoming. During the autumn of 1986 a simple card-type database software was introduced. Four databases were generated in the hope of something more sophisticated becoming available. These were: an edited version of the context record; a finds record; an record of inscribed biographical information; and a photographic archive. These databases, while primitive, enabled the checking of errors and the production of crude statistics during the first phase of post- excavation.

Two further priorities had to be met: first, a means of producing more than 50,000 items of information in numeric form for an archive which could then be used by students of mortuary archaeology anywhere was needed; second, a means of drawing information from each of the databases so that instant inventories could be generated, checked, and assessed for inference quality (IQ) was highly desirable.

A relational data management system cloned from Dbase III+ was purchased during the summer of 1987. The databases were transferred into this system and a single data catalogue then generated containing a comprehensive burial catalogue. This catalogue has been produced as a series of fiches here.

It was intended that a recording system suited to rapid retrieval of pro-forma data onto a microcomputer database be designed eventually, so that information from similar sites could in future be retrieved at a small fraction of the time and cost spent at Christ Church. In view of a number of recent vault clearances which have been carried out commercially because of fears of another large-scale project like that at Christ Church, this seems an important goal.

3 The archaeology of Christ Church

'. . . it is not really difficult to construct a series of inferences, each dependent upon its predecessor and each simple in itself. If, after doing so, one simply knocks out all the central inferences and presents one's audience with the starting-point and the conclusion, one may produce a startling, though possibly a meretricious, effect.'

(Sherlock Holmes, in Sir Arthur Conan Doyle's *The Dancing Men*)

1 Introduction

Analysis of the burial areas at Christ Church has been based on the division of the crypt into sections within which a discrete sequence of events can be reconstructed. Within these sections the two-letter codes used to identify areas during excavation have been retained. Figure 3.1 shows the location of the areas with their codes. Area names which use surnames of individuals do not necessarily imply ownership of any vault, only the presence of at least one individual with that family name. The northern parochial area was given its name because it was situated north of the main parochial vault, not because it was considered to be a parochial, that is public, vault.

All contexts were analysed to assess, firstly, whether the deposit was primary or secondary (for a discussion of the criteria used and their implications see chapter 7); secondly, where that context fitted in the depositional sequence. All burial contexts have *termini post quem* which have been included in the Burial catalogue. The Burial catalogue, in conjunction with the structural matrix and the stacking plans in this chapter, should enable a more or less complete reconstruction of the burial sequence along contrasting hypothetical lines. Alternative hypotheses could also be generated with reference to the enclosed data.

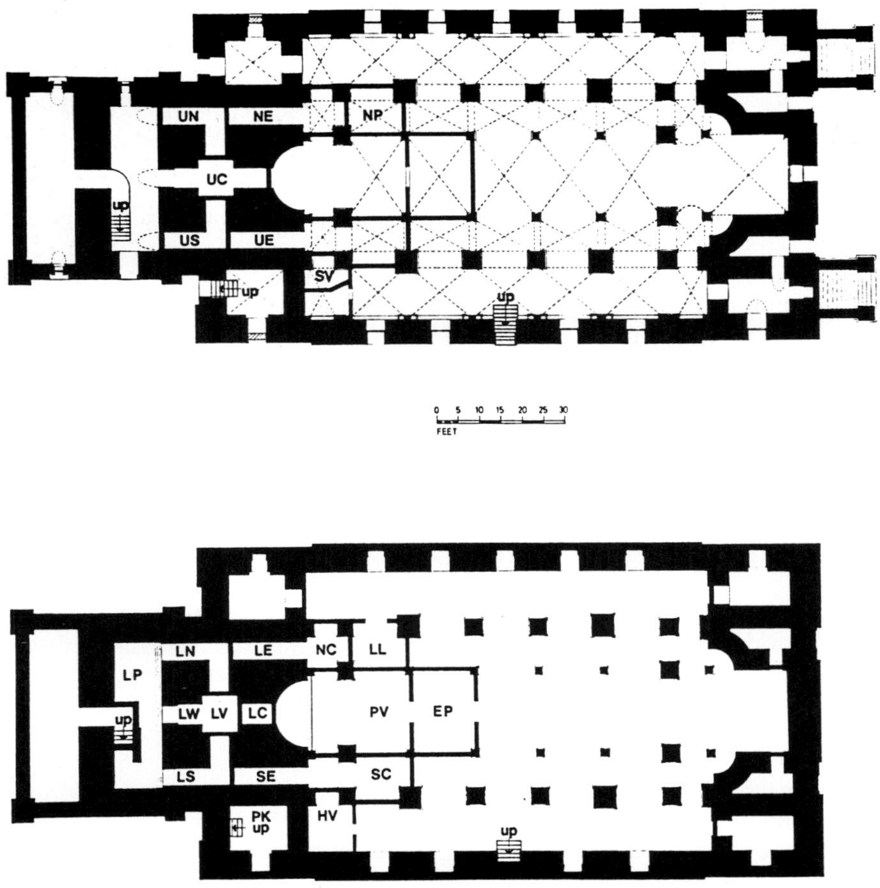

Figure 3.1 Split-level plan of the excavated areas with area codes

The following sections describe, as far as possible, sequences reconstructed on the basis of use of the law of superposition, together with inferences which we have drawn from evidence of formation processes. The limitations of this approach, and an assessment of the difficulties and advantages of identifying process traits at Christ Church, have been discussed fully in chapter 7. In this chapter deposits of refuse have been discussed according to their materials and the sequence in which they have been deposited. In chapter 7 these deposits have been assessed for their process traits and been discussed accordingly. Statistical summaries are provided at the beginning of each section. These follow the form used by Boddington (1987), indicating the number of observations and the sample size and expressing the former as a percentage of the latter. For example, a vault containing 100 interments of which 10 were primary would be expressed as 10/100/10.0%.

Figures are presented for each area which indicate the relative positions of burials and the number of the lowest burial in a stack. These figures can be located on the larger projections of the excavated areas. Figure 3.2 shows a plan of the lower west areas which can be located on the larger reconstructions. Interments have been shown diagrammatically, differentiating between adult and infant burials, and horizontal and vertical positions. Other features have been shown where they bear on the sequence reconstruction. The burial outlines are representative, and are not intended to indicate precise dimensions of interments, only relative locations to facilitate interpretation.

2 The lower west areas

Summary

The lower west area comprises the lower portico, lower south-west tunnel, lower north-west tunnel, Lemaistre\Pontardant vault, and lower central west tunnel. They occupy an area of approximately 44 m^2. Access to this part of the crypt was from the south-west corner of the church beneath the stairs leading up to the portico at the west end. A doorway leads to a series of steps down to the lower level.

The lower north, lower south, and Lemaistre\Pontardant areas were used as private family interment areas between 1755 and 1849 or later. By the end of this period these areas, which are likely originally to have been open to friends and relatives of the deceased, seem to have fallen into a state of some untidiness: after 1849 most of the burial activity was occurring in the part of the crypt accessible from another entrance to the east. The area to the west of the portico, beneath the steps at the west front, may have been used for interments, and was large enough to have been used as a public vault. If it was so used, it was cleared, probably after 1867 when it was mentioned in a Queen's Council ruling (see Appendix A4), and the interments placed in the lower portico area and sealed up with a brick wall. Rearrangement of interments seems to have been a regular feature in this area as elsewhere. A large box containing several corpses of which some had undergone post-mortem experiments had been deposited in the area, along with three bodies which seem to have been disposed of hastily at the end of the area's period of use.

Lemaistre\Pontardant area (LV)

The area contained 21 interments:

Adults	15/21/71.5
Infants	06/21/28.5
Primary	14/21/66.7
Secondary	07/21/33.3
Dated	15/21/71.5

The area is vaulted from the four corners at 0.80 m above the floor to a maximum height of 1.55 m. Four tunnels lead off. These are vaulted with a maximum height of 1.2 m, as are all the lower tunnels. In form, these are all of the 'railway arch' type described by Litten (1985) and typical of the early 18th century. The entrances to the north, east, and south tunnels were blocked by brickwork, apparently prior to the introduction of interments. The area is constructed of brick, unplastered, with a floor of exposed brick, in common with all the lower west areas. At the centre of the vault a circular depression which corresponds to the vertical axis of the steeple appears to have been employed during construction as a marker for a plumb line. A mortar deposit which covered most of the floor of this vault is associated with the construction of the vault. Access to the area during the burial phase was from the west via the lower central west tunnel (LW).

No records have been found relating to the purchase of the area as a private vault, nor do any of the inscribed slabs indicate such a purchase. However, all the surviving biographical details point to an exclusive occupation by the Lemaistre\Pontardant family. Surviving interment dates range from 1755 to 1795.

The first primary dated interment (2204) was placed against the north wall in or after 1761. The only dated coffin prior to this date, an infant's interment of 1755 (2249), was located above a coffin (2206) in the north-west corner dated after 1768, and must, therefore, have been relocated. Interments may have been introduced prior to 1755, but it would seem that around 1768 part of the area was rearranged, and three deposits of broken coffin wood created at two opposing corners and against the east wall. Of these, the deposit in the south-east corner contained two adult coffin sides, textile fragments and a sherd of an 18th or 19th century storage jar, together with a few items of coffin furniture and a small quantity of building or demolition debris. An infant's coffin (2250), dated to 1760, was placed adjacent to 2206 (its breastplate was located in the deposit of coffin wood underneath), and another

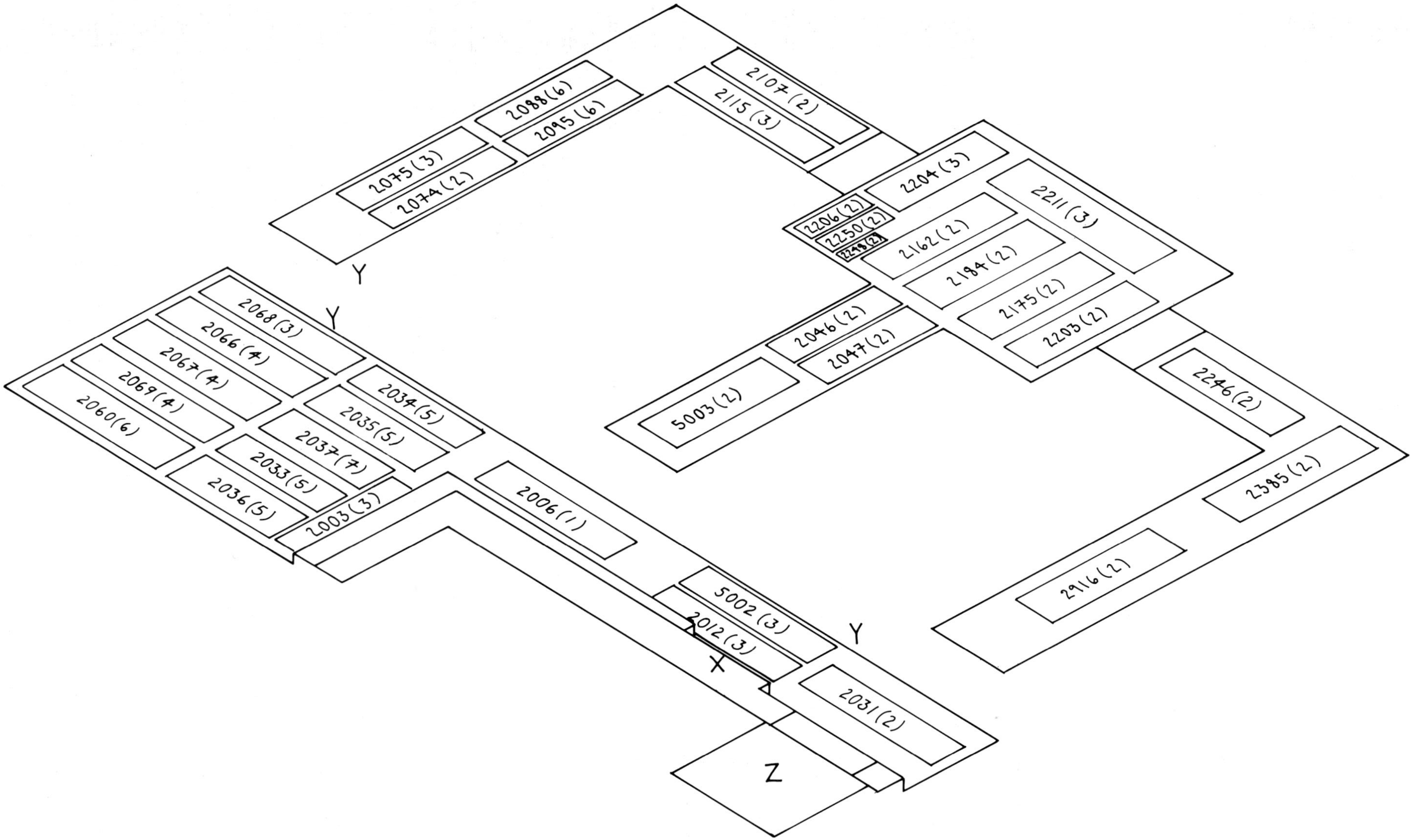

Figure 3.2 The lower west areas: burial scheme. Reading left to right: LP, LS, LV, LW, LN

(2188), dated 1763, was placed on top. The next primary interment (2211) is of 1779 or after. It was placed against the east wall on two coffin supports, overlying one of the post-1768 refuse deposits. Above this were two further adult interments (2251, 2207), both primary, of 1784 and 1790 respectively.

Another adult interment of 1779 (2162) was placed adjacent to 2211 to the west, probably two months or so later. Interments of 1781 (2070) and 1791 (2063) respectively were placed on top of 2204 and 2162, and a third, undated, placed on 2070. Two further infants' interments were placed adjacent to those in the north-west corner between the 1768 rearrangement and 1793 or after (see Fig 3.3). The remainder of the area was apparently filled during the period 1791–5, in three stacks, almost certainly from the southern wall north towards the access point from the LW area. Some rearrangement must have taken place during this period, since interments 2175 and 2184, dated to 1777 and 1794 respectively, were placed on the floor between 2203 and the west tunnel after 1795. Two further interments appear to be secondary. After the introduction of 2052 access to the area would not have been possible.

Lower central west tunnel area (LW)

The area contained four interments:

Adults	04/04/100
Infants	00/04/000
Primary	00/04/000
Secondary	04/04/100
Dated	00/04/000

The area consists of a single tunnel, vaulted, aligned east–west, 1.2 m high, running from the lower portico area (LP) to the west entrance into the LV area. It was constructed in the same way as the latter. At the west end the tunnel is truncated by the spreader courses of the tower foundation, where the floor drops 0.58 m. The four interments, all adults, were placed at the east end of the tunnel. None have surviving dates. 2046 and 2047 overlie a breastplate belonging to 2175, from the LV area and dated to 1775. The four interments must have been placed in the tunnel after 1795, and all show signs of having been moved from elsewhere. West of these coffins a wooden box, 1.72 m x 0.52 m x 0.55 m (labelled 5003 on Fig 3.2) lay with its western end truncated at the tunnel entrance. Figure 3.4 shows the box during excavation. The box contained an inner wooden casket. No lid remained. The state of the iron brackets which must have secured the lid originally suggested that the lid had been removed in a somewhat hamfisted manner. The box contained the partial remains of at least 21 individuals, some of which must have been articulated at the time of burial. A few of the skulls showed signs of post-mortem surgery, including trepanation (Fig 3.5). It has been suggested that the contents of this box may represent the remnants of scientific exper-

Figure 3.3 Interior of LV, north-west corner (LV:0155.02)

imentation of the sort provided for by the 'resurrection men' of the 18th and 19th century (see chapter 4). The box had been overfilled with these remains, presumably after the removal of the lid; it is likely, therefore, that the deposit represents more than one incidence of this type of disposal. The door was not sealed; its original door was removed before 1867.

Lower south-west tunnel (LS)

This area contained six interments:

Adults	06/06/100
Infants	00/06/000
Primary	05/06/83.3
Secondary	01/06/16.7
Dated	05/06/83.3

The lower south-west area comprises two arms of a tunnel set at right-angles to each other formed by the blocking of the south entrance to the LV area and the blocking, also by brick, of half of the lower south tunnel. The tunnels are 1.2 m high and of the same 'railway arch' construction as the LW area. Access was from the west via the LP area. Here also the tunnel is truncated by the tower foundation. There are no surviving records relating to ownership. The first interment (2246) was placed in the north arm of the area after 1798, and appears to be primary. Above this was an undated interment, of which the coffin had collapsed or decayed *in situ*. The last four interments belonged to individuals from the same family. The first of these (2949), dated to 1797, was placed at the east end of the

Figure 3.4 Wooden box 0051 during excavation

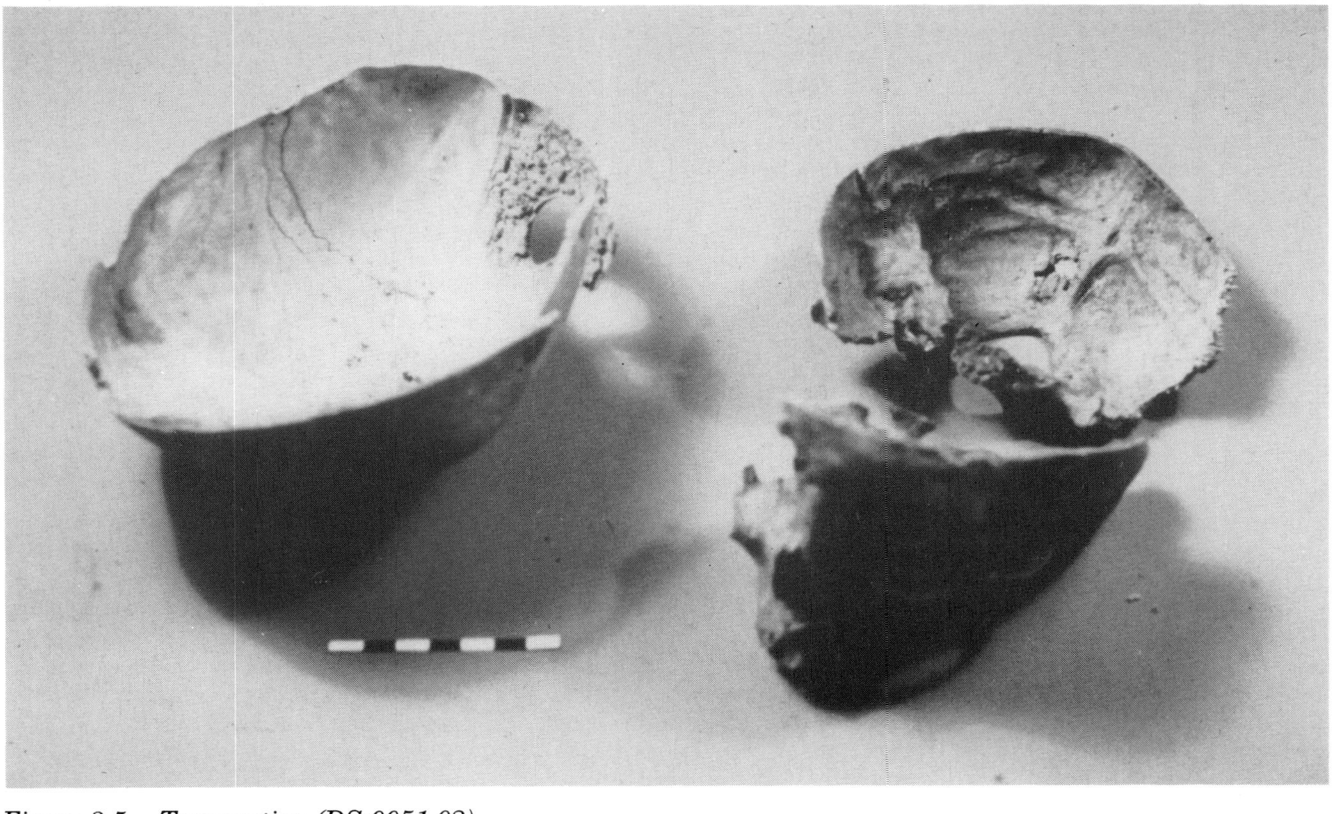

Figure 3.5 Trepanation (DS:0051.02)

Figure 3.6 Lower north-west tunnel (LN:0000.03)

east–west tunnel, after 1798, and is therefore a secondary deposition. The other three interments appear to have been primary depositions after this date, of 1806, 1847, and 1849. The entrance was not sealed, but the original door was removed.

Lower north-west tunnel area (LN)

The area contained 22 interments:

Adults	20/22/90.9
Infants	02/22/09.1
Primary	04/22/18.2
Secondary	18/22/81.8
Dated	03/22/13.6

This area is of the same dimensions as the LS area, and of the same construction, with its shorter arm running south. Access was from the west end of the east–west tunnel, and again, there were no surviving records of ownership. The first interment was placed in the southern arm. Of the first two one was undated and the other (2115), was a primary interment of 1781. Above this was another primary interment of 1794, adjacent to one of 1798. The final interment in this arm was an infant, undated, which appears to have been brought from elsewhere in the crypt. After this point some maintenance or deliberate disturbance is evinced by a layer of coffin-wood dust overlying these interments. The activity responsible for this seems to have been the introduction of a quantity of disarticulated human bones and skulls and decayed or decaying coffin wood, which almost certainly had been subject to attack by boring insects. At the east end eight skulls had been grouped together. Below this deposit, but later than the interments in the southern arm, several interments had been placed. Three of these (2093, 2092, 2090) had been brought in, still articulated, on coffin bases. Other individuals and parts of individuals had been piled upon these, in various states of articulation and completeness (see Fig 3.6). It is inferred that these deposits represent the residue of a maintenance process which occurred in another area.

Subsequent movement in this area is represented by two interments without coffins (2081, 2084) at the west end, and five additional interments, none dated, in coffins. All of these appear to be secondary. Finally, at a date probably some considerable time after 1798, and after the deposition of the first of the interments which filled the LP area, a deposit of rubble, bones, and funerary and non-funerary debris was made from the west end, extending at least 2 m into the tunnel, and then a young adult interment was placed on top at the west end. It is extremely difficult to demonstrate the exact nature of this later activity, and its date may lie anywhere between 1798 and 1867. Resolution of this problem is obscure because of the nature of the LP area environment, which was very damp and caused severe destructive changes (for possible reasons, see below, LP), rendering the inference potential of a crucial sequence relatively low. The tunnel was not sealed, but the original door was removed.

Lower portico area (LP)

The area contained 62 interments:

Adults	61/62/98.4
Infants	01/62/01.6
Primary	05/62/08.1
Secondary	57/62/91.9
Dated	08/62/12.9

The lower portico area is formed between the west edge of the tower foundations and, to the east, the foundation wall for the portico at the west front of the church. The steps from the south-west corner of the church lead down to this area. It appears that this entrance into the crypt may have been intended to allow access for friends and descendants of the deceased. An iron railing bounded the top of the stairwell, and an iron bar, which could have held a padlock, would have restricted access. The ceiling is vaulted from a height of 1.35 m to a maximum of

Figure 3.7 Stacks at north end of LP

1.76 m, forming five bays at the point where the west wall of the tower foundation starts.

At the base of the steps an arch, which at some point held a door frame, led west into the area beneath the steps at the west front of the church. It was thought that this area, which was used as the excavation workshop, had never contained interments, but this hypothesis has recently been re-examined in the light of the sequence reconstructed in the lower portico area (see below). The wall which was constructed to block the interments in the LP area from the steps has not been excavated to its foundations, but it seems likely that when the area was being used for interment it was not in existence. It is likely that until at least the 1850s the doors hung to enclose the tower foundations during the construction phase were still used as entrances to the north and south tunnels and the Lemaistre/Pontardant vault. An inscription set into the wall enclosing the interments, relating to an interment of 1791, seems originally to have been set into the east face of the wall forming the stairwell (marked X on Fig 3.2). Fixtures set into the west wall of the tower foundation (marked Y in Fig 3.2) may have held wreaths or candles.

It seems likely that the interments in the LP area are all of the same phase: at least 90% of them are demonstrably secondary deposits, and the rest are probably also secondary. If this is the case then interment did not take place until after 1849, when the latest interment was deposited into the lower south tunnel. The nature of the material dumped over the area suggests that the interment of 62 individuals and the dumping of the sealing deposits may be associated with the 1867 abandonment phase.

A sequence may be reconstructed as follows, although it must be stressed that the scope for interpretation being low in this area means that a wide range of other possibilities may be considered. At the extreme south of the LP area, beneath the steps, a small area (marked Z in Fig 3.2), enclosed by a door, contained a neat pile of coffin debris overlying a deposit of destruction debris. This coffin debris may be associated with the interments lying in the greater part of the lower north tunnel, which were conspicuously without whole containers. More coffin debris was deposited at the west end of the lower south tunnel.

This appears to signal the beginning of an orchestrated plan for the consolidation and abandonment of the area as an interment facility. The doors leading to the three tunnels, and affixed during construction, were removed. The exposed ends of the north and central tunnels were filled with debris, which may be associated with the widening of

Figure 3.8 Rubble against blocking wall of LP

Commercial Street during the 1840s and 1850s, but includes bricks, assorted demolition rubble, sherds of domestic ceramics, and leather scraps, similar to deposits elsewhere in the LP area and the rest of the crypt which date from around 1866–7.

At the north end of the LP area discrete deposits of mixed rubble and coffin debris may indicate some tidying of the area. There seem to have been two different types of use being made of the area at the same time. Coffins were being introduced into neat stacks at the north end (see Fig 3.7), while at the south demolition rubble was being introduced, partly blocking area Z containing the stack of coffin wood, to a height of 0.65 m or so. The space above this dump was blocked by a brick wall laid directly on the dump. The dump subsequently subsided by about 0.2 m, leaving the base of the wall suspended. A further sequence of coffins interspersed with rubble and coffin debris appears to have occurred towards the south end at the same time as, or after, the introduction of the last of the coffins at the north end.

There is no evidence of the location from which these 60 or so interments were brought. However, if, as seems likely, this activity is associated with the Queen's Council ruling of 1857 (see Appendix A4), we may conjecture that they had originally occupied the area directly beneath the main church steps, to the west of the LP area. The original floor of this area has never been examined. The present floor, which is at a level approximately 0.2 m above that of the LP area, is made up of a mixed concretion similar to that which sealed the Peck vault (PK) and the Leschallas vault (LL). This concretion was laid after the construction of the wall which sealed the LP area. Limescale deposits which cover the walls and ceiling of this area indicate that it has been subject to leakage of water over a long period. If interments had originally been deposited in here it might be expected that they would have suffered from damp and fungal rot, as had those which were redeposited in the LP area, which shows no signs of having suffered from excessive damp. Indeed, the three lower tunnels indicate that conditions were very dry during the period of primary interment.

The deposits of coffin debris which occur frequently in the LP area are good candidates for a secondary deposition of coffin wood gathered from such a clearance procedure. Since the area beneath the west steps was and remains open to the streets via two windows, it may have been the case that the sealing of the entrance arch was not considered adequate under the terms of the Queen's Council ruling of 1859 (see Appendix A4), and that all those interments which had been deposited there had to be moved. If the secondary interments in the LP area had been introduced from elsewhere in the crypt then the large deposits of secondary refuse from this area would have to be explained by activities for which there appears to be no evidence, quite apart from considerations of public decency which would have resulted from the carriage of 60 or more half-rotten coffins from one side of the churchyard to another. The covering of the floor of the western extension with concrete might be accounted for by an attempt to lessen the effects of the extremely damp conditions there. After the beginning of the construction of the blocking wall a large quantity of rubble was introduced into the LP area, apparently to seal the entire area (Fig 3.8). Three interments, all articulated bodies, but none contained in coffins, were placed just over the north face of the wall and, concealed from sight, were bricked up with the other redeposited interments. Figure 3.9 shows the blocking wall of the LP area immediately prior to excavation.

3 The upper west areas

Summary

Figure 3.10 shows burials within these areas. They encompass the western halves of the upper north and south tunnels, together with the upper central passage as far as the entrance to the main parochial

Figure 3.9 Blocking wall of LP prior to excavation

area: a total of approximately 27.5 m^2. In design they follow the pattern of the lower west areas. The square, vaulted chamber within the central passage corresponds to the LV area below. The tunnels are wider than those at the lower level, with ceilings up to 2 m in height; they are of the same type and phase of construction.

The upper north-west areas were used for interment between 1729 and around 1816. Many coffins were relocated during this period, and although the three vaults may have been exclusively private before the beginning of the 19th century, they appear to have been filled with interments from more than one family afterwards. The upper central area appears to have been cleared of interments around 1816, allowing access into the parochial areas. The demolition deposits which covered the parochial and south-eastern areas must have been introduced through the upper central areas in or around 1867. The area between the north and south walls of the

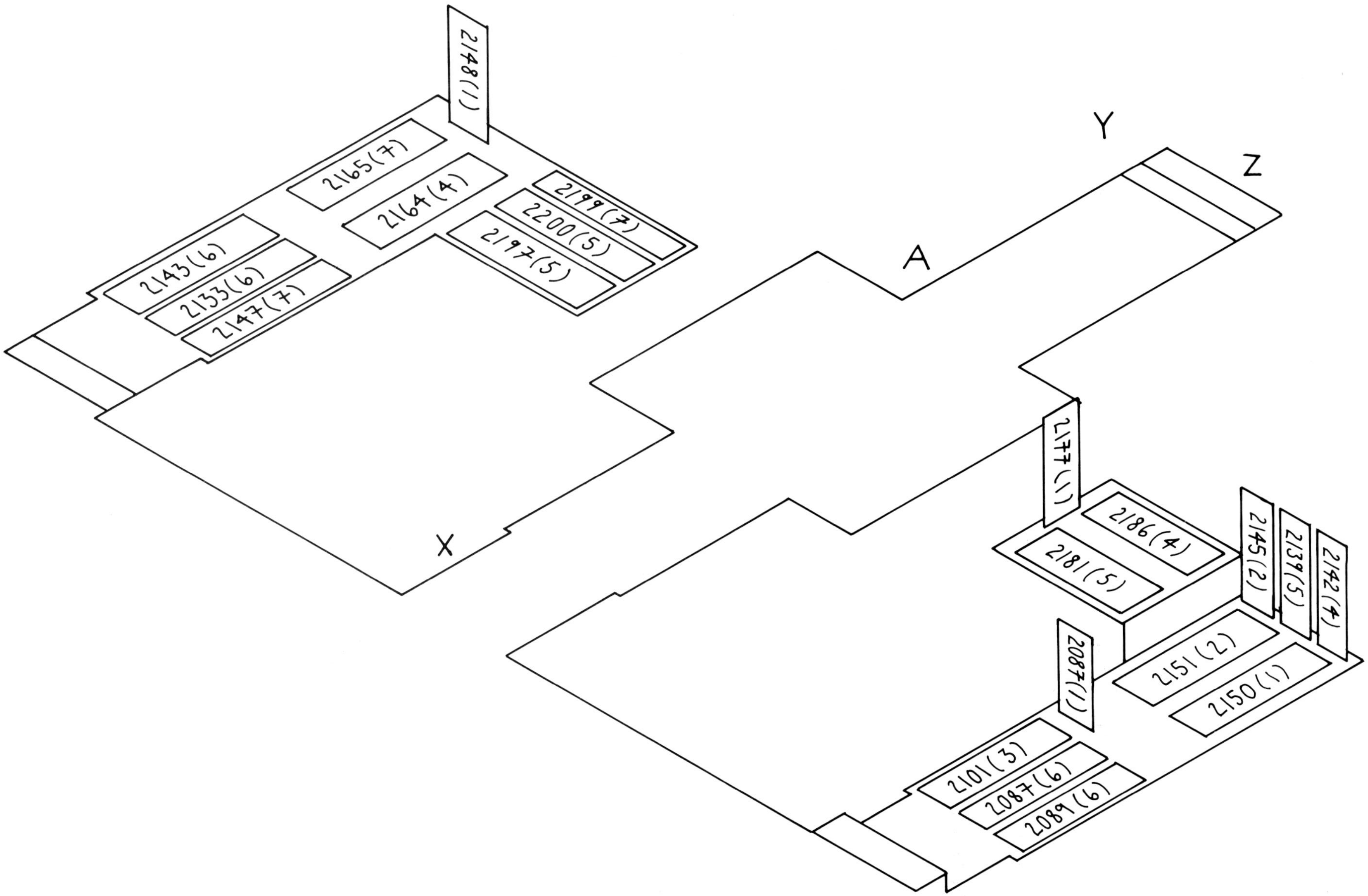

Figure 3.10 *The upper west areas: burial scheme. Reading left to right: UN, UC, US*

Figure 3.11 View of UN before excavation (UN:0000.01)

portico foundations, north of the crypt entrance and staircase, was probably used as a storage area or vestibule and remained accessible from the earliest interment period up until the beginning of the excavations in 1984.

The upper north-west tunnel (UN)

The area contained 45 interments:

Adults	34/45/75.6
Infants	11/45/24.4
Primary	12/45/26.7
Secondary	33/45/73.3
Dated	15/45/33.3

This area follows the shape of the lower north-west area below, with a long western arm and a shorter southern arm. At the east it is truncated by a blocking wall, beyond which lies the upper north-east area (NE). At the west end a narrow doorway provided access from the south-west doorway of the crypt. An inscription (Appendix C20) indicates that the vault may have been used by the Vernezobre family. However, there are a minimum of 13 families represented in the 45 interments, and therefore the likelihood of an exclusive family occupation such as that which seems to have existed in the LV area is remote. Interment dates in this area range from 1802 to 1816.

If the Vernezobre inscription does indeed refer to this vault, then it belongs to a period, before 1800, for which we have no evidence in this area: no Vernezobre interments were identified during excavation. The first deposits for which there is archaeological evidence are of sawdust with occasional human bones, at the extreme east end of the area, and on the west side of the south arm. These appeared to have dropped from coffins, which commonly contained sawdust in order to prevent movement of the corpse (see chapter 5). These deposits overlay two coffin supports which were re-employed for later coffins. The first three coffins placed in the vault, across the south arm, were secondary. Above the centre one of these was a primary interment of 1802 (2189). One of the lowermost interments (2197) was contained in a badly disturbed wooden coffin. The lid was missing, as was the head end. Coffin debris had been deposited between the coffin and the wall of the vault on the west side. At some point the skull was removed from, or fell out of, the coffin. It may be that a skull (2170) located at the top of this stack belonged to this interment. There were five other interments in the southern arm which may have been primary, dating between 1813 and 1816. The others, dated between 1810 and 1815, had undergone some considerable disturbance. An undated primary interment (2168) at the top of one of these stacks had two thin sheets of lead covering the top half of the body.

The other interments in this area seem to follow a similar pattern of occasional primary interment and frequent rearrangement of coffins. The remaining coffins are arranged in two stacks, one at the east end of the vault north of the southern arm, and one to the west of this towards the entrance to the vault (Fig 3.10). The presence of two iron hinges set into the southern wall of the vault, where the brick foundation abuts the stone arch at the entrance, with a corresponding iron catch set in the north wall, suggests that a door was originally fixed at this point. It was removed before the last stack of coffins was introduced. Some coffins appeared to have been moved more than once, and some (eg 2200) had certainly been brought from elsewhere in the crypt, where rendered walls had left deposits of plaster adhering to coffin upholstery during transit (and possibly from the upper central area — see below). Two of the three young adult interments lying on the floor at the west end of the area (2138, 2141) had originally been stored upright, one with the head facing down, as evinced by the location of sawdust deposits inside their coffins. After three of the interments at the west end had been deposited a layer of rubble, including bricks and plaster, was deposited over them, before the introduction of more coffins. This later activity may have taken place at any time after 1816 and before 1867, and the area may have undergone further 'maintenance' during that period. In 1867 a deposit of demolition rubble derived from the main body of the church itself was introduced into the area, down the whole length of the western arm of the tunnel. After this event the area was sealed with a blocking wall.

The upper south-west tunnel (US)

The area contained 40 interments:

Adults	35/40/87.5
Infants	05/40/12.5
Primary	07/40/17.5
Secondary	33/40/82.5
Dated	12/40/30

The construction of this area is similar to that of the upper north area. However, the floor drops 0.2 m at the west end and the north arm is consequently raised by 0.2 m. At the east end the tunnel is truncated by a blocking wall. At the west end a doorway enclosed the area on the exterior of the blocking wall.

No inscription survives for this vault. At least eleven families are represented. The date range is 1729–1809. The earliest evidence for movement in this area is trampled mortar deposits on the north side of the west arm. These deposits represent mortar which had been dropped when wet and had set *in situ*. Clean sand deposits occur above this, and on the floor of the north arm. Above the sand in the north arm were located five coffin supports which had been made with reused wood from door frames. The first primary dated interment, lying on these supports, is of 1729 (2186), on the east side of the north arm. Adjacent to this is a primary interment of 1744. More primary interments of 174–? (2169) and another post-1744 were added to these. A deposit containing two broken candle holders, some broken items of coffin furniture, human bones, and other assorted debris overlay these interments. Above this deposit were laid two further primary interments of 1750. Of these first six primary interments, five belonged to the same family. It may be the case that the area was during this period being used as a private vault.

Above these interments were located four others, all secondary, including a young adult interment (2177) which was upright. Of these two were dated, from 1759 and 1792. This may suggest that interment became more infrequent after 1750. One of the secondary interments (2158) was in a coffin containing sawdust packing, and had possibly been in a poor state prior to its relocation as a deposit of sawdust of the same type was found on the north side of the west arm about halfway along the tunnel, possibly indicating the previous location of 2158.

The next interments in the vault were three adults placed at the east end of the tunnel. None are dated, and two are probably secondary. On top of these interments were placed eleven upright coffins, probably all secondary, dated between 1798 and 1809. Of these, 2145, 2134, 2130, 2132, 2128, 2123, 2119, and 2118 showed signs of having originally been interred in a horizontal position (Fig 3.12). A deposit of demolition debris was located above these coffins, followed by coffin debris.

West of these interments were three stacks of horizontal secondary interments, up to six high, with a further upright interment (2087). These must all have been deposited after 1809. Many of them lay on their sides.

At some stage during the burial period a deposit of decayed coffin wood was left near the entrance. This deposit had been transformed by trampling and kicking into a corner, reflecting further cultural activity.

Finally, the door must have been removed, the entrance partially blocked, and then a quantity of debris consistent with the 1867 abandonment phase introduced before the wall was completed.

The upper central tunnel (UC)

This area contained no interments. It is formed by a tunnel running west to east from the west edge of the tower foundation to the apse of the main parochial area (Fig 3.13). At the centre a square area corresponding to the Lemaistre/Pontardant area

Figure 3.12 Detail of US showing upright coffins (US:0000.01)

directly beneath forms at present an unenclosed chamber. The walls of the west tunnel and the central chamber had been whitewashed, though unrendered. The east tunnel had been plastered and repainted on more than one occasion. There is no direct evidence that the upper central area ever contained burials. However, a door originally stood at the west end, opening inwards, outside which, at the point marked X on Figure 3.10, an inscribed slab (Appendix C16) indicates that the area may have been a private vault. A sequence of events in this area has been reconstructed within which it has been inferred that there were interments at here at one time. This reconstruction is based partly on information which has come to light since the excavations finished in September 1986.

The floor of the area was created in part by the vaulting of the area underneath. The floor of the central chamber was formed by the ceiling of the LV area, made up with bricks in each corner to form a

Figure 3.13 UC before removal of blocked eastern arch

more or less level surface. This surface was covered with a bed of sand approximately 0.05 m thick, and a brick floor was laid on top. In the east tunnel, at the point marked Y on Figure 3.10, two bars, perhaps of iron, were put up, one above the other, across the tunnel at chest height. It seems likely that these were meant to prevent workers or onlookers from falling into the space now occupied by the parochial area during the construction phase. These bars were later removed. The east tunnel was plastered at the same time as the parochial area, and the door fitted at the west end. The east end of the east tunnel was blocked by a wall (marked Z on Fig 3.10) which con tained a false arch that would have been visible from the parochial area. This alteration was probably the result of a decision to 'sell' the area as a private vault. This is likely to date from around 1741/2, when the inscribed slab records the vault as having been purchased (Appendix C6).

During this period the east tunnel was 0.2 m lower than at present, dropping at a comparable point to the upper north-east tunnel. At the time of excavation the tunnel was level with the floor of the central chamber, and composed of a similar concretion to that below the portico steps, and above the Leschallas and Peck vaults. This concretion was not removed during the excavation. However, during recent work in the crypt it has been shown that it covered a deposit of rubble and coffin debris.

There is no surviving evidence that any interments had been placed in the upper central area. However, a series of events points to the likelihood that this was the case. After the closure of the main parochial area in 1813, a further interment was made there, for reasons which are not apparent. The interment (2860), apparently primary, must have been introduced from the west end of the parochial area (PV) through the east wall of the the east tunnel of UC. This wall was partly dismantled and truncated at a height of perhaps 0.5 m above the lowered floor of the east tunnel. If any interments remained in the upper central area they must have been moved at this time in order for access to be made into the parochial area. If such a clearance occurred in 1816 it would coincide with the introduction of several secondary interments in the upper north west area (see above) about this time. Any debris remaining in this area seems to have been cleared to the east tunnel, and a wall was then erected at a point marked A in Figure 3.10. The existence of this wall was only demonstrated after the end of excavation when layers of paint were carefully removed to reveal the scoring lines and mortar traces from the construction of the wall. After 1816 it seems likely therefore that this area was free of interments. The presence of many nails and brackets set into the wall of the area may indicate that it was used for storage, or as a workshop area.

The blocking wall erected in or after 1816, possibly using bricks from the floor of the central chamber which had been pulled up, probably stood until around 1867. This passage remained the only feasible entrance into the parochial areas and the south-east areas. It was through here, then, that the 200 or so tonnes of rubble and other debris were introduced into these areas to seal them. The 1816 wall was pulled down, and the bricks perhaps used as a ramp up into the parochial area. These deposits, which represent the abandonment of this area as a burial place, were introduced, and then the upper central area blocked off again, this time at the extreme east end where the original wall had been. The east tunnel was levelled off at the same height as the rest of the area and sealed with concrete.

4 The north-east areas

Summary

Figure 3.14 shows the upper north-east tunnel (NE), the lower north-east tunnel (LE), the north chasm (NC), the north parochial area (NP), and the Leschallas area (LL). The sequence in these areas is closely interlinked. They cover a total area of 26.5 m^2. They were accessible from either the north arch of the north chasm or the modified doorway of the Leschallas vault before it was converted to private use. The latter seems more likely, since a lamination of trampled material similar to that in the PV (see chapter 7) was located on the floor of the LL area. This indicates that the area had been subject to more traffic than can be accounted for by its sole use as a private area.

Interment in these areas seems to have been infrequent before the abandonment phase, when the Leschallas burials were covered and the LL and NC areas used for secondary interment. The two north-east tunnels may have been family vaults.

The upper north-east area (NE)

The area contained 25 interments:

Adults	24/25/96.0
Infants	01/25/04.0
Primary	07/25/28.0
Secondary	18/25/72.0
Dated	03/25/12.0

This area is a continuation to the east of the upper north-west area, although the floor drops by 0.2 m immediately to the east of the blocking wall. At the east end the tunnel had originally been separated from the north chasm by a wooden door. Such a door still survived at the end of the upper south tunnel prior to excavation. The floor of the vault is some 1.66 m above the floor of the north chasm.

Access to this vault was from the east via the north entrance to the north chasm (marked X on Fig 3.14), or more likely from the doorway into the LL area. Interment in this tunnel seems to have been infrequent; it may have belonged to the LeFevre family, for whom there is an inscribed slab in the church (see Appendix C1), since two of the three dated interments belonged to that family. Dates range between 1746 and 1823.

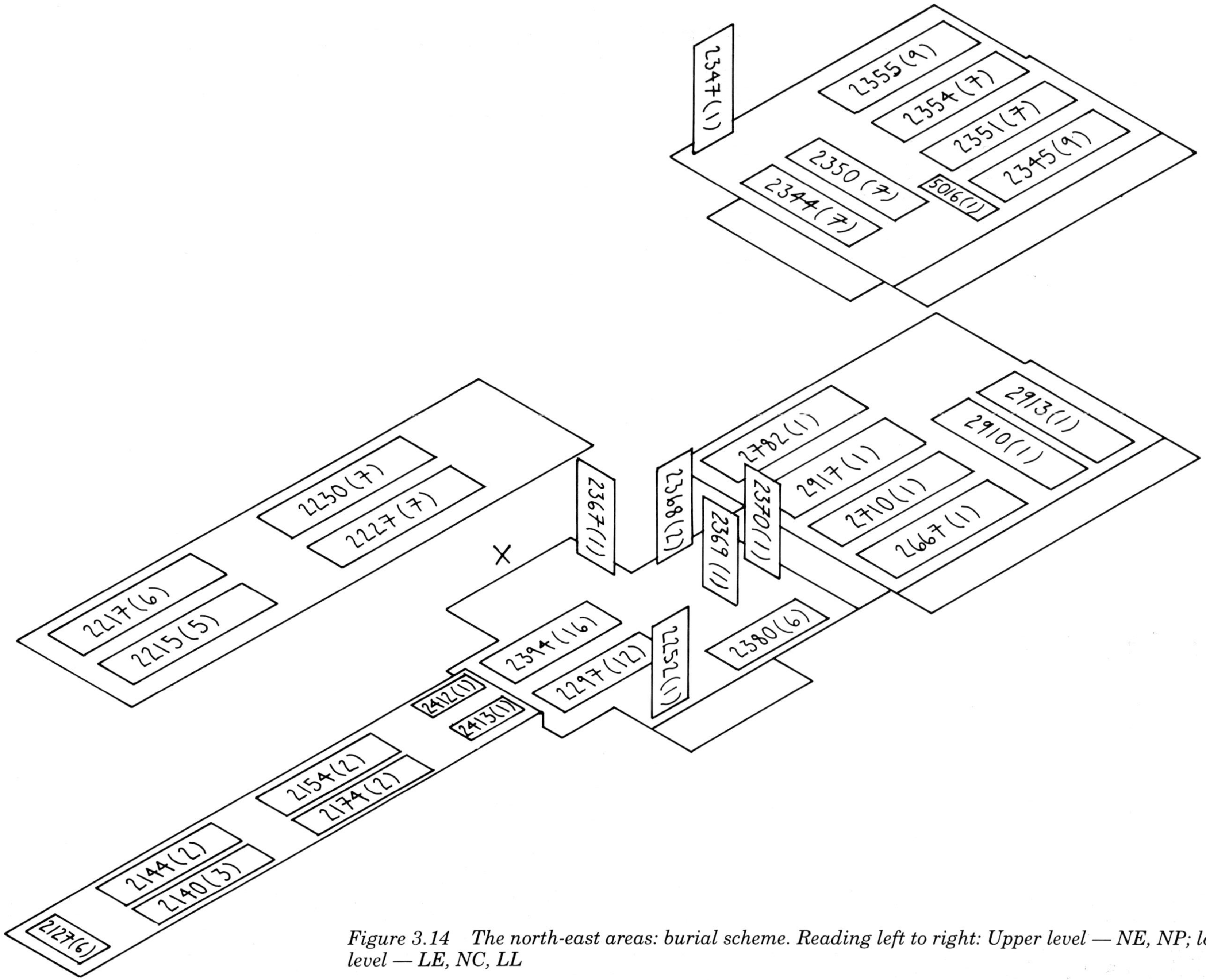

Figure 3.14 The north-east areas: burial scheme. Reading left to right: Upper level — NE, NP; lower level — LE, NC, LL

Five primary interments were made at the west end of the tunnel, four on the north stack and a single coffin on the south stack, before a deposit of coffin debris was introduced over the west half. Six further interments, including two infants, were introduced at the west end, although these all appear to have been secondary. The uppermost three of these were in a dilapidated state and may have been introduced after the creation of the two coffin stacks to the east (Fig 3.15). These latter, of which all but one interment were secondary, include two interments dated 1808 and 1823. Various interments had either bones or parts of the coffin missing.

The lower north-east area (LE)

This area contained 17 interments:

Adults	06/17/35.3
Infants	11/17/64.7
Primary	03/17/17.6
Secondary	14/17/82.4
Dated	06/17/35.3

This area is a continuation of the lower north-west tunnel. At the west end it is truncated by a blocking wall, and at the east by the north chasm. It was probably enclosed by a wooden door in the same manner as the upper north tunnel, within the same frame. Access was from the north side of the north chasm or through the LL area. The vault may have belonged to the Chevalier/Chauvet family, of which there are three dated interments towards the west end. Dates range from 1752 to 1845, although it is probable that the later interments were secondary. An inscription on the inside of the north wall of the crypt refers to the Vaux family (Appendix C8), of which two members were relocated in the LE area.

The first deposits in this tunnel are of dust, animal bones, and a hazelnut shell. These are inferred to have been the result of animal action (faunal turbation). Six infant interments, all secondary, were introduced at the west end of the tunnel. None of them bore dates. The first primary dated interment was an adult introduced east of these in 1752. This lies above 2140, a primary but undated interment. North of these two interments a primary interment followed by another of 1757 were introduced. Both of the dated interments belonged to the same family, together with an infant of 1754 lying above the 1752 interment. East of these two secondary adult interments were introduced.

At the east end of the tunnel a deposit of fine ash and soot occurred between the introduction of these and two further secondary interments, one of 1845. At some point a deposit of coffin debris was introduced between the two west stacks and the east stacks, possibly associated with the secondary deposition of several of the coffins, and two infants were placed at the entrance to the tunnel. These are dated 1799 and 1811, but appear to have been deposited here after 1845. Finally coffin debris, consisting mainly of coffin sides, was slotted between the edges of these two infant interments and the walls of the tunnel. This activity may be associated with the abandonment phase depositions in the NC area (see below).

The Leschallas area (LL)

The area contained 6 interments:

Adults	06/06/100
Infants	00/06/00
Primary	06/06/100
Secondary	00/06/00
Dated	06/06/100

Figure 3.14 shows the positions of burials in this area. The area was defined by four brick walls to which access was gained via a doorway in the northern wall. The doorway had been remodelled more than once, but no evidence was ever found to explain these alterations. The area was vaulted to a height of 3.84 m. On the north side of the northern wall a metal inscription had been affixed to the wall, over the doorway, with the name Leschallas inscribed on it. It

Figure 3.15 NE, looking east into NC (NE:0467.01)

is therefore inferred that the vault was bought in its entirety by the Leschallas family, and indeed four of the six interments in the vault are of the Leschallas family. Dates range from 1829 to 1852.

All six interments were placed on the floor of the vault (Fig 3.16), with that of the earliest date located at the furthest point from the doorway. There is no evidence of any movement of the interments or any other activity in this area. After 1852, when the latest interment was introduced, the wooden frame around the doorway and the door was removed and the space bricked up. At some point after this the western wall to the vault was partly dismantled and truncated to a height of 1.31 m; a dump of building material rubble was introduced into the Leschallas area up to the height of the truncated wall and sealed with a concrete, similar to that in UC, the area west of LP, and that sealing the PK area. As the rubble contained fragments of debris associated with the 1866 restoration programme it is inferred that this second sealing of the Leschallas vault was part of the 1867 abandonment process.

The northern parochial vault (NP)

This vault contained 48 interments:

Adults	45/48/93.8
Infants	03/48/06.2
Primary	00/48/00.0
Secondary	48/48/100
Dated	11/48/23.0

This vault is defined by the upper parts of the north, east, and south walls of the LL area, as its floor is the upper side of the concrete which was introduced to seal the LL area in about 1867. The western limit was defined as the western limit of that concrete. Since all the interments in the area were secondary and they appear to have been introduced in the planned abandonment of 1867 it is not thought that the area was intended to be a private disposal area. None of the named individuals were related to one another, so it could be that they were introduced into the NP area from a parochial area or several private areas. The dates range from 1799 to 1843.

The first activity in this area after the introduction of the floor must have taken place after 1852, being the latest date in the Leschallas vault, and probably occurred after 1867. Twenty secondary interments were deposited in the vault in the first phase of deposition, up to four in height, and across the whole of the floor area of the vault. Coffin debris was introduced during the same period between interments. Most of this debris comprised whole sides or lids of coffins. Subsequently a deposit of ash, charcoal, bricks, and items of coffin furniture, and a fragment of an inscribed slab, was made over all of these interments, and continued into the north chasm (NC) area to the west.

Further secondary interments were then introduced above this debris, but following the same stacking arrangement which had been used initially. This phase is represented by nineteen interments and

Figure 3.16 Burials in LL (LL:0000.01)

Figure 3.17 Upper part of the NP from the PV area (NP:0000.01)

further discard of coffin debris. Of these six contained no identifiable human remains. This phase was sealed in the same way as the previous group of interments: a deposit of similar materials including coffin debris and fragments of an inscription relating to the Bowden family, who were identified from the EP, LP, and UN areas. This deposit did not appear to continue into the north chasm. The last nine interments in the vault, all secondary, were introduced after this debris, along with several deposits of coffin debris. A lidless infant's coffin containing no identifiable remains was excavated from this phase. The lid and inscription belonging to this interment were retrieved from the second-phase dump below. This supports an inference that although there are three discrete phases of activity in this area, they are all part of a single strategy of maintenance and rearrangement consistent with what is known of the 1867 abandonment phase, and probably occurred within a short time of each other (ie within the year 1866–7). An axonometric reconstruction of the interments in this area, together with some discussion of the implications of the apparently 'reversed' stratification, appeared in an article by Adams & Reeve (1987, 249). Figure 3.17 shows the uppermost part of the vault as it was revealed from the parochial vault during excavation.

The north chasm (NC)

The area contained 41 interments:

Adults	40/41/97.6
Infants	01/41/02.4
Primary	02/41/04.9
Secondary	39/41/95.1
Dated	05/41/12.2

This area was defined on the west by the tower foundation pierced by the upper and lower north tunnels, on the south by a blocking wall separating it from the parochial area (PV), on the east by the truncated blocking wall of the NP area, and on the north by a remodelled entrance within a blocking wall which now separates it from the entrance passage to the alcoholic vagrancy centre. Dated interments range from 1812 to 1847.

Burials are shown in Figure 3.14. The north entrance was almost certainly used for the introduction of the interments in the two north-west tunnel areas and the NP area after the conversion of the LL area into a private vault. Prior to 1829 access was probably from the east via the LL area. The floor of this area was covered with a thin layer of sooty ash similar to that found in the lower north-west tunnel, and dates from after 1801, and possibly from around 1836 (see chapter 1). It may have been at this time that the doors at the east ends of the tunnels, together with the single frame which supported them during the construction phase, were removed. An oak beam lying in a slot at the base of the foundation at this point remained *in situ* during the interment phase (Fig 3.18). If any interments occurred here prior to 1867 they were rearranged at that time to allow for access to the NP area, and all of the excavated deposits appear to have been generated during the abandonment phase. The first recognizable activity is represented by the destruction of the wall blocking the north entrance. This first blocking probably took place after 1823, when the last primary dated interment (2221) was placed in the upper north-east tunnel (NE). When the wall was taken down, probably in 1867, over-zealous destruction seems to have been responsible for pulling up some of the brick floor near the doorway. The void created was filled with coffin debris, which was trampled down and then made up level with the rest of the floor with demolition debris.

Figure 3.18 West end of NC showing remains of beam under burial (NC:0862.01)

Figure 3.19 Upright coffins in NC (NC:0806.01)

The first interment phase took place after the first interment phase in the NP area, and is represented by the introduction of ten secondary interments: two stacked horizontally in the extreme south-east corner, two placed upright against the entrance to the NP area (Fig 3.19), and two horizontal stacks of three coffins adjacent to the entrance to the lower of the tunnels (LE). A deposit of demolition debris similar to that ending the first phase in the NP area was introduced, not in order to cover these interments, but filling the spaces between them. It may be inferred that the purpose of introducing this deposit was to discard the material rather than to seal the interments. The inclusion of fragments of plaster mouldings from the main church interior suggests a provenance and date for this deposit (ie the renovation of 1866–7).

Coffin wood debris was then introduced between two of the stacks and the entrance to the LE tunnel, and into the south-west corner of the area. At this point it is possible that an undated interment was placed upright in the south-west corner (2252). An

Figure 3.20 NC prior to excavation (NC:0000.01)

infant interment (2247) contained in an iron coffin (see chapter 5) was placed on a shelf created by the springing of the arch across the south blocking wall. It bore no stratigraphic relationship to any other interments and was not dated, and could therefore have been introduced at almost any time, though more likely later than earlier.

Six further secondary interments were added to the stacks created during the first phase, two aligned north–south, corresponding with the second phase in the NP area. In the same period of activity an upright coffin (2368) was introduced against the north pier, another south-east of the entrance, and a third lying at about 45 degrees between two of the upright interments against the entrance to the NP area. These were added to by a deposit of coffin debris. In this phase one undated interment (2357) may have been primary.

Five adult interments were deposited against the west end. The remaining fourteen interments were placed in two stacks at the west end, resting partially on the floor at the east end of the upper north-east tunnel (NE). One of them (2232) may have been a primary interment: it was in a substantially better condition than those around it. One coffin contained two sets of remains of which one (2238) was undisturbed; the other (2239) was disarticulated and appeared to have been added to the coffin some time after the original body. Above this coffin lay the base of another which supported two further unenclosed individuals. The last deposits in the area prior to the reblocking of the north entrance were of coffin debris, in considerable quantity. There was a sealing deposit of demolition rubble immediately to the south of the entrance. This was introduced, as in other areas, after the blocking wall had been constructed to a considerable height. Figure 3.20 shows the vault immediately prior to excavation.

5 The south-east areas

Summary

Figure 3.21 shows schematically the interments in these areas. They constitute an area of approximately 24 m^2. Access to all these areas was from the east end of the south chasm (SC) during the interment phase. To the north the area is separated from the main parochial area by two blocking walls; it is bounded on the west by the tower foundation pierced by the upper and lower south-east tunnels; to the south a blocking wall separates it from the SV and HV areas. Fig 3.22 shows the area under excavation.

The south-east areas were accessible from an entrance in the south wall of the church which still exists (see Fig 1.6). Primary activity is most closely associated with the use of the upper tunnel as a

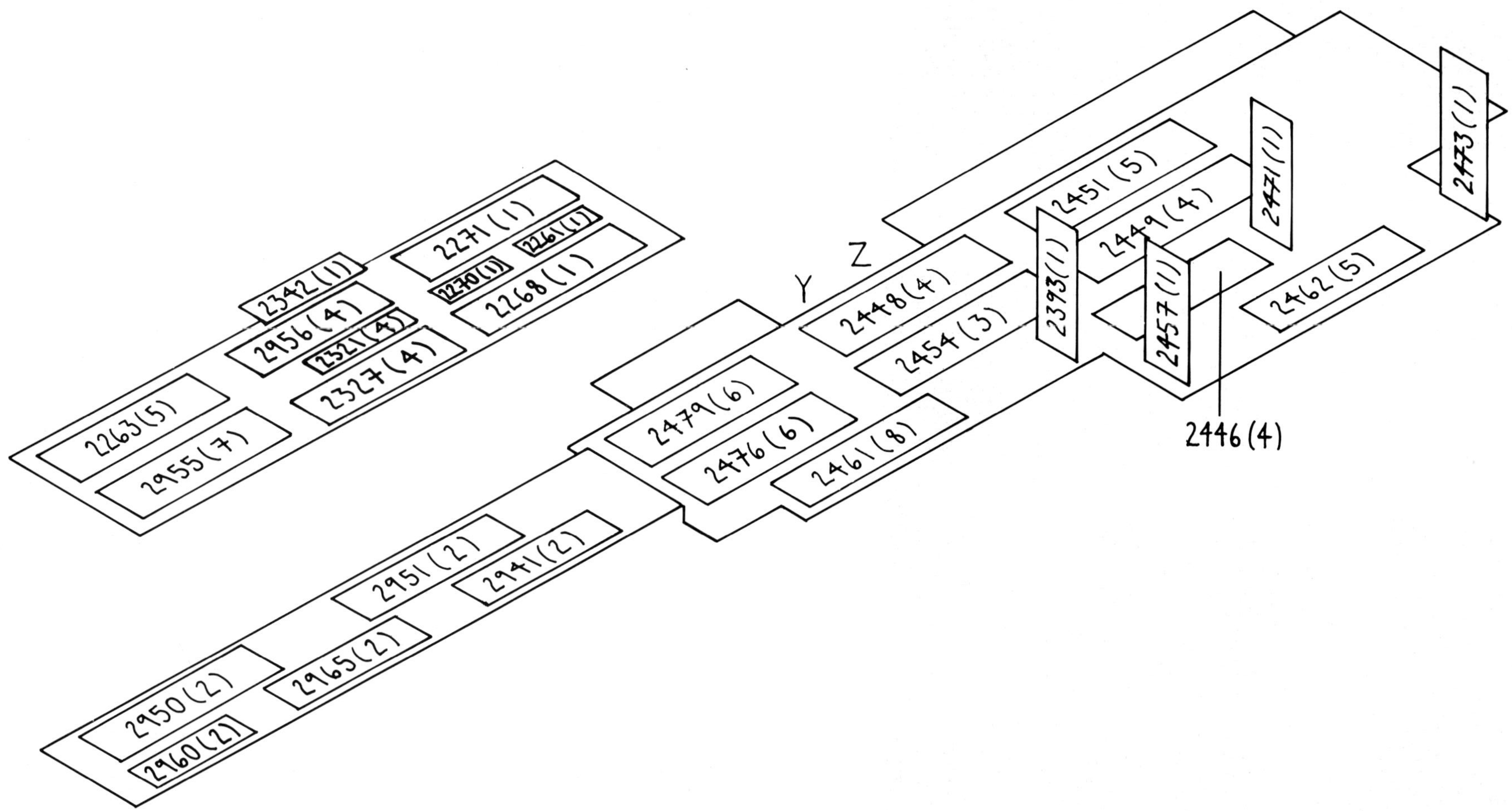

Figure 3.21 *The south-east areas: burial scheme. Reading left to right: upper level — UE; lower level — SE, SC*

Figure 3.22 SC during excavation, from the east (SC:0000.01)

private vault between 1731 and 1801. After that time primary interment was sporadic, although the south chasm was used as access to the Simpson vault (SV) until 1843. The filling of the area can be identified with the abandonment phase between 1858 and 1867. It may have been the first area to be filled as part of that phase.

The upper south-east tunnel (UE)

This area contained 29 interments:

Adults	21/29/72.4
Infants	08/29/27.6
Primary	17/29/58.6
Secondary	12/29/41.4
Dated	13/29/44.8

An inscription on the south facing side of the wall (marked 12 on Fig 1.5) indicates the presence in the south-east areas of the vault of the Mesman family (Appendix C12). This is most easily identified with the upper south-east tunnel. Of the 16 interments named on the inscription, 8 were positively identified in this area, and a further 7 infants for whom coffin plates do not survive probably correspond with the 7 infants named on the inscription. One member of the family was interred at the extreme south-west corner of the parochial area. Five other interments in this area were not mentioned on the inscription, but belonged to the same family. The first named interment on the inscription, of 1731, was not located in the vault.

Ten interments, nine of them dated, form the initial sequence in the vault. These were placed in four stacks at the west end of the tunnel, between 1732 and 1794, and all are primary. After 1793 a secondary interment, undated, was added to the south-east of these stacks, and a coffin lid was deposited on the stack to the north, after 1794. Above this a secondary adult interment was deposited, and five adult and two infant interments deposited on the two south stacks at the west end. Three of the adult interments were dated 1765, 1769, and 1801, and were secondary. Three infant interments were placed above each other in the centre, and may be primary. Above these a primary adult interment was deposited. A further primary adult interment was deposited across the two eastern of the four existing stacks. Subsequently a deposit of coffin debris and human bones was introduced, covering this interment.

A layer of ash with some soot, similar to that in the north-east areas, covered all the deposits up to this point, and lay over the floor at the east end of the vault. This event occurred after a secondary interment of 1801 in this area but may have been around 1836 (see chapter 1). Following this a large quantity of coffin debris was introduced at the west end, covering the westernmost stacks and lying between the stacks to the east.

Subsequent interments in this vault were all secondary, and none was dated. They were deposited no more than one high at the east end of the tunnel. The two adults and three infants, in coffins which had been severely disturbed, were introduced, probably in 1867, after a layer of demolition debris covering the interments in the south chasm had spilled into the tunnel. The door frame still existed at this time, but it is unlikely that the door was still in place during the abandonment phase.

The lower south-east tunnel (SE)

The area contained ten interments:

Adults	08/10/80.0
Infants	02/10/20.0
Primary	07/10/70.0
Secondary	03/10/30.0
Dated	00/10/00.0

There is no surviving evidence to show whether this vault was used as a private burial area. No inscriptions appear to relate to it, and no biographical details survive from any of the interments.

Of the ten interments seven may be primary, although bone survival was very poor in most cases. The lack of dated material means that elucidation of a sequence here is virtually impossible. After the introduction of the interments a deposit of silt and assorted debris including three human skulls was made at the east end, and may be associated with the abandonment phase. Above this (see Fig 2.3) a deposit of coffin debris was introduced at the entrance.

The south chasm (SC)

The area contained 48 interments:

Adults	47/48/97.9
Infants	01/48/02.1
Primary	16/48/33.3
Secondary	32/48/66.7
Dated	06/48/12.5

It is thought that access to this area was from the east, since the parochial area was blocked off by partition walls from as early as 1729. There seem not to have been any interments in this area before at least the turn of the 18th century. Fixtures for inscribed slabs remain on the north side of the area marked Y and Z on Figure 3.21. These are likely to have referred to the Mesman vault in the upper south-east tunnel, which was in use at least until 1801. The slabs were eventually removed in a somewhat crude fashion, damaging one of the piers slightly. The slab now on the south side of the southern wall of this area may be one of the two formerly on the north wall, or a copy of one or both of them. Interment probably did not take place in this area until after 1843 when an inscription slab was set into the wall blocking the Simpson area (SV) from the south chasm. It appears that an ashy deposit noted elsewhere, which may be inferred to have been introduced around 1836, underlay all other deposits in this area (see chapter 7). A wall existed running north–south across the centre of the SC area prior to the construction of the south wall. There is no evidence of its purpose; it probably dates from before 1801.

Two interments dated to 1839 and 1852 (2476, 2461) appear to have been primary depositions towards the west end of the area. However, circumstantial evidence indicates that interment in this area may have formed part of an early phase of abandonment before the 1866–7 closures. The insertion of the Simpson slab into the blocking wall of that area suggests that the south chasm was accessible to visitors after 1843 and possibly much later (see below, section 6). The interment dated 1839 was located on the floor below the inscription, and if indeed it was a primary deposit its position would seem to have overstepped the boundaries of propriety that seem to have prevailed at Christ Church.

Other circumstantial evidence relates to the nature of deposition in this area. It is the only area of the crypt in which large amounts of refuse material were consistently used to cover layers of coffins before the introduction of subsequent interments (Fig

Figure 3.23 General view of SC from the east (SC:0000.04)

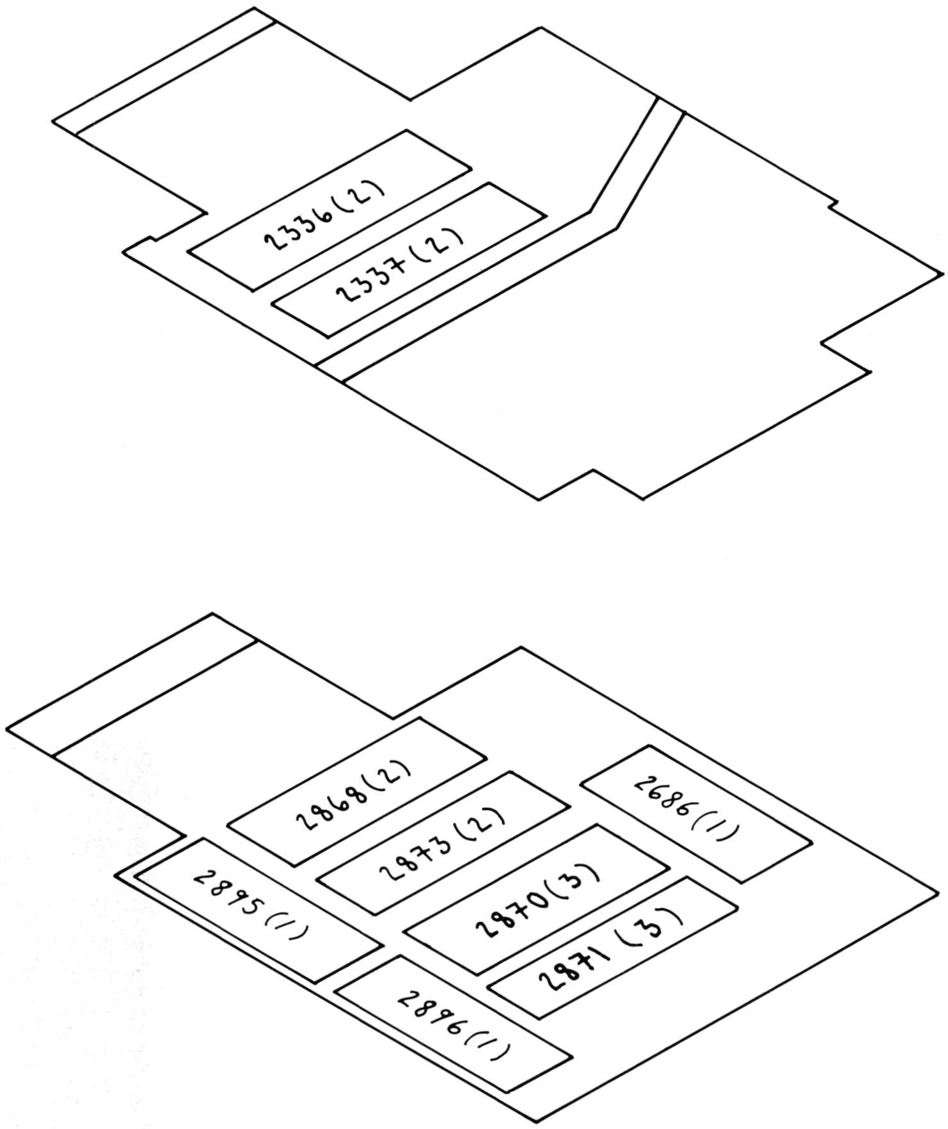

Figure 3.24 The Hebert\Simpson areas: burial scheme, lower and upper vaults

3.22). The refuse deposits themselves are unlike any found elsewhere in the crypt: they contain a high proportion of humus, and what appears to be a garden soil-type matrix of pebbles and topsoil together with oyster shell and ceramic tiles. It is possible that these deposits derive from work being carried out in the churchyard in 1859, which extended it west by some 15–20 m to its present position fronting on Commercial Street. This would date the deposition phase here to after the 1858 prohibition of vault burials at Christ Church (Sheppard 1957, 168). If this was indeed the case, then all the interments are likely in any case to be secondary, especially since they are virtually all wooden caskets without lead inner caskets (see below in PV and in chapter 5).

The first interments were deposited at the west end of the area. Five adults were placed in three piles immediately to the east of the lower south-east tunnel. A deposit of building refuse was made at the southernmost point of the area, and four interments placed across the width of the area, leaving a gap of some 2 m between them and the coffins at the west end. The *terminus post quem* for this activity is 1852, from 2461. The first dump was then introduced, covering all these interments and extending east as far as the end of the coffins.

The next phase is represented by the introduction of nineteen interments, including one upright (2457), over the dump, and a small deposit of coffin debris and human bones adjacent to it. Two new stacks were created in the space between the earlier interments south of the pier at Y. Following this another deposit of apparently similar soil was made over these interments.

The third phase of interment is represented by fifteen coffins added variously to existing, though covered, stacks. At the west end these were now six high, although in places coffins were stacked as little as four high. One coffin contained the remains of four individuals, two adults (2390, 2391) and two infants (2392, 2393). Two upright interments were then deposited before the introduction of the final dump covered all the exposed interments. This final dump rests against the east blocking wall of the area which, like others in the crypt, seems to have been half constructed before the deposit was completed. Part of the brick floor at the east end, which had been removed, may have been employed in the construction of this wall. The entire depositional phase in this area may well have been completed within less than a year, judging by the homogeneity of the deposits.

The final phase of activity in the south-east area occurred during the major abandonment phase of 1866–7. The top 1.5 m of the wall between the pier at Y and the east end was dismantled and a deposit of several tonnes of the typical refuse encountered throughout the rest of the crypt was introduced, filling the area to within about 1 m of the ceiling. This must have been brought through the upper central tunnel. On top of this deposit, at the entrance to the upper south-east tunnel, the four interments mentioned in the description of the upper south-east tunnel were introduced. Immediately south of the dismantled wall a deposit of charcoal had apparently been tipped over the wall from the parochial area.

6 The Hebert\Simpson areas

Summary

A single, roughly square area was divided by a brick vault and floor into two separate areas for burial, totalling 23.5 m^2. Figure 3.24 shows these areas. The original purpose of the area has not been definitely identified, but it seems certain that its use as a burial facility was not intended during the original construction phase. The upper of the two vaults was created by a brick floor vaulted from beneath and sprung from a height of 1.4 m. This insertion was clearly not part of the original construction phase, and certainly occurred after the area had been abandoned from its intended use. Figure 3.25 shows the entrances to both vaults from the north.

The Hebert area (HV)

The area contained 13 interments:

Adults	13/13/100
Infants	00/13/100
Primary	08/13/61.5
Secondary	05/13/38.5
Dated	00/13/100

This area contained no identified individuals. However, two breastplates were recovered bearing the surname Hebert, and it is possible that the vault was purchased by a single family, although no inscribed slab survives.

The floor of the vault consists of limestone slabs closely laid and levelled. It is approximately 0.1 m higher than the average of the other, brick floors in the crypt. There are two flues which rise from the south wall of the vault. One of these, at the west corner, rises into the south aisle of the church, and

Figure 3.25 SV and HV areas from the PV area (HV:0000.08)

would originally have led beneath the wooden floor of the main body of the church. This flue corresponds with flues in the other bays, both north and south, of the church. It is likely (A D Mason, pers comm) that these features are the remains of the hypocaust system designed by Hawksmoor to heat the church, but never fully used for that purpose. The archaeological evidence suggests that the Hebert area may have been the boiler room (the second flue appears to have held a stovepipe). If so, then the almost total lack of a draught in this part of the crypt may offer an explanation for the apparent inefficiency of the system. No remains of the boiler itself were recovered during excavation.

The north entrance to this area was apparently never used for introducing deposits. The wall blocking it from the SC area was built at the same time as the vaulting was constructed. An entrance into the Peck vault (PK) was blocked before any deposits occurred. Access for the introduction of burials was from the east through a doorway, which rose from the floor up to the vaulted ceiling. Some material was also introduced from above via the two flues.

The first deposit in the area was introduced from the west flue: a pile of bricks which had been dumped from above and had fanned out from the south-west corner (Fig 3.26). Two deposits of charcoal mixed with ash were then introduced through the two flues into the south-west and south-east corners. It appears that although attempts were periodically made to block these flues by pouring cement and plaster down them from above these were thwarted by water which drained down them either from the church or the exterior. Transformation of some deposits indicated that at times the water must have been pouring down the flues in considerable quantity. The next identifiable deposit occurred at the north end of the area. Coffin debris and a few human bones were placed or dumped where the area narrows between two piers on either side.

Since there are no dated interments it is difficult to determine when primary burial deposits began; it is insufficient to date interment by the presence of dated breastplates (of 1788 and 1794), which may or may not have belonged to the interments in this area. These were introduced as part of deposits of coffin debris.

The first interments were secondary. Two wooden coffin supports were placed north–south, about 1 m apart, on the floor. One of these was perhaps 2 m in length. Several deposits of coffin debris were made on top of these. A secondary interment was then deposited towards the north end of the area above the earlier deposit of coffin debris. Against the west wall a deposit of human bones was made. This included two identifiable individuals (2895, 2896) which had still been articulated at the time of deposition, and bones from more than one other individual. Above these, in the north-west corner was a further deposit of coffin debris and human bones.

Three interments which may have been primary were placed above debris in the centre of the area lying east–west. Above these more coffin debris was introduced. Five interments were added above these lying east–west from the north end towards the south wall. After another dump of coffin debris had been introduced at the north end the final interment was placed near the east wall, lying north–south in front of the entrance.

The only datable event of this sequence is the introduction of a dump identified with the abandonment phase of 1866–7, which was introduced from the east after the partial blocking of the eastern entrance. The dump did not extend across the entire area. In place of the charcoal used extensively elsewhere in the crypt a layer of cinders and burnt wood, possibly coffin wood, was introduced above this dump. The deposit included items of coffin furniture which had been subject to fire. It appears that this deposit was introduced very shortly after it had been burnt, and was indeed still very hot, because the bones of the individuals interred without coffins against the west wall had been charred in the process.

Figure 3.26 Early dump through flue in the south-west corner of HV (HV:1786.01)

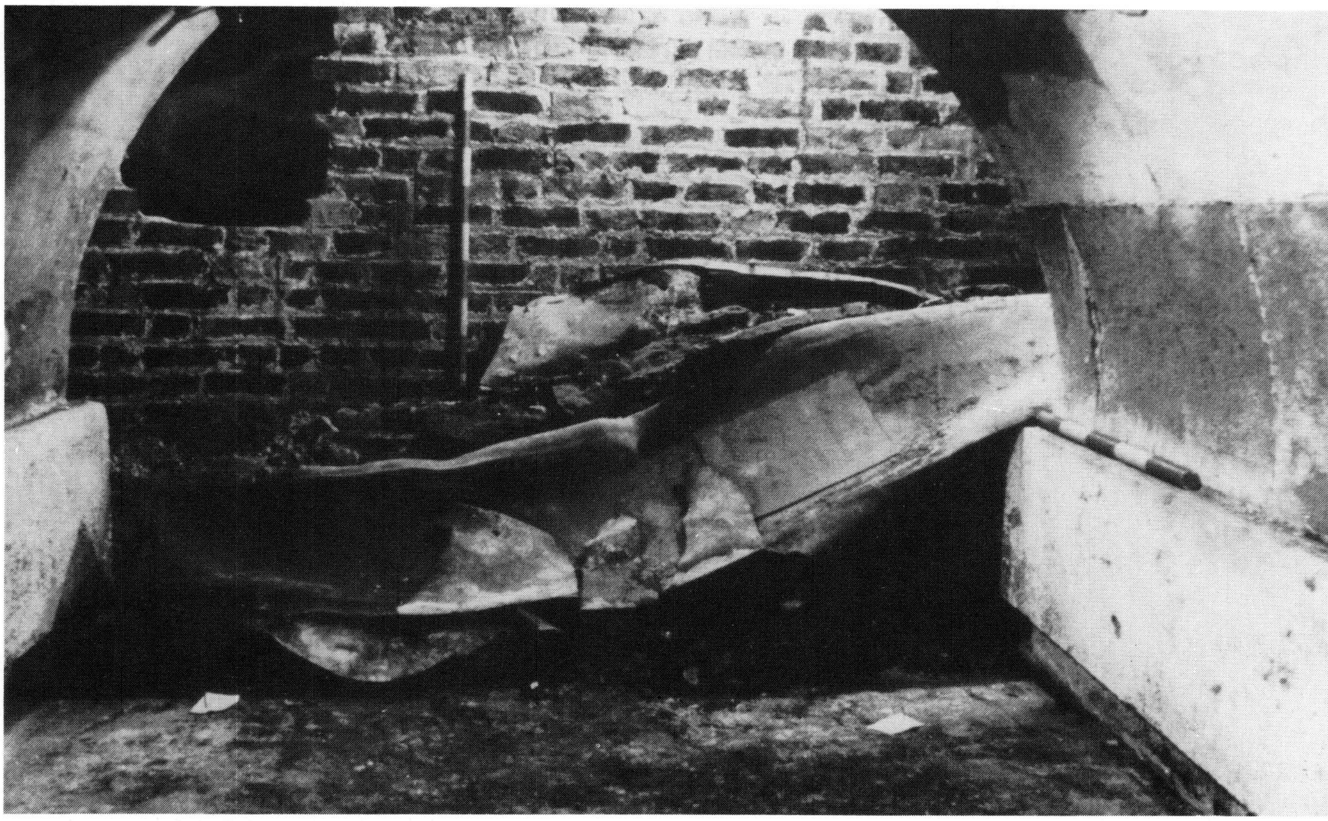

Figure 3.27 Burials in SV from the north (SV.0000.01)

The final event in the vault, after the completion of the east blocking wall, was the introduction of a deposit of plaster through the eastern flue. This plaster appears to have been carried by a gush of water which entered with such force that it disturbed the southern end of interment 2686, which had lain under the abandonment dump. Subsequent transformations of the deposit depended largely upon the moisture content in the area, probably as a direct result of water entering from the flues.

The Simpson area (SV)

The area contained 4 interments:

Adults	04/04/100
Infants	00/04/000
Primary	00/04/000
Secondary	04/04/100
Dated	04/04/100

The Simpson area was created by the horizontal division of the vault originally intended as a boiler room. An arch leading to the PK area was blocked. The wall blocking the HV area from the SC area did not originally rise above the floor of the Simpson area. This entrance remained accessible until at least 1843, and perhaps after (see discussion of SC area). An entrance on the east side also seems to have been used until after interment had ceased here.

The floor of the area was covered with the same sooty ash deposit which is present elsewhere in the crypt and may be dated to 1836 (see chapter 1). This underlay the wall built to divide the Simpson area in half. It seems likely that the area was used for private burial for the Simpson family. Two of the four interments bear the name. None of the interments are primary, but it is likely that they were interred as primary deposits between 1767 and 1843 in the south or centre of the vault and then moved to the northern half to accommodate the wall which was built across the vault (Fig 3.27).

No satisfactory explanation has been advanced for the construction of this wall. It provides for access from the east to the southern half which, when excavated, contained two inscribed slabs (Appendix C22 and C9). These may have been placed there during the 1960s when a hole was made in the east wall for the purpose of examining the state of the vaults. However, the accessible area after 1843 or thereabouts was restricted to this southern half, no more than 3 m. The answer may lie in the opening at the south end onto the churchyard. These may have been blocked following an order in Queen's Council of 1859 (Appendix A4). It is possible that prior to this order offensive smells were emanating from the Simpson area and that the wall was created to prevent this, and perhaps also to interrupt the view of decaying bodies.

The movement of the interments to the north end of the vault created debris which was piled on top of

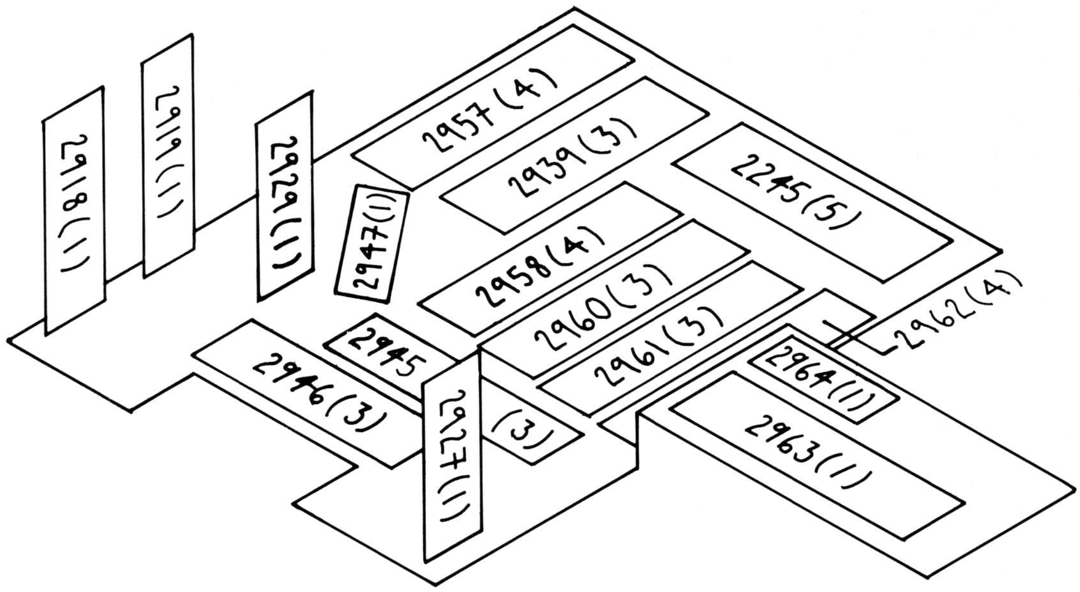

Figure 3.28 The Peck vault: burial scheme

the interments. After 1843 the vault was sealed by a brick wall from the north, into which an inscribed slab was set (Appendix C25).

7 The Peck area (PK)

The area contained 39 interments:

Adults	37/39/94.9
Infants	02/39/05.1
Primary	34/39/87.2
Secondary	05/39/12.8
Dated	13/39/33.8

The Peck area (see Fig 3.28) forms an almost independent adjunct to the main burial area of the crypt. Entered by a separate entrance on the east side, it appears to have been the first area of the crypt specifically set aside for interment when it was promised to Edward Peck in 1727 (see above, chapter 1 and Appendix C2). It measures 14 m^2. Figure 3.29 shows the interior of the vault looking west during excavation.

On the east side archways leading to the HV and SV areas were blocked, probably during the phase which converted the vaults for burial. On the south side a recess opened onto the churchyard, and below it a short tunnel provided a recess for interments. The stone steps which lead into the area from the west descend to 1.2 m above the brick floor and are set on a brick plinth. It is unclear why the steps do not continue to the floor, but it may be that they were constructed before the floor was lowered in the 1720s, if indeed this happened. Access to the floor of the vault may have been accomplished by the use of additional wooden steps which may have been portable and possibly were used elsewhere in the crypt, for example to gain access to the upper east tunnels and the Simpson area.

The range of dates for primary interments in this area is from 1749 to 1847, though burial probably occurred here from at least 1730. The sequences of maintenance and rearrangement so familiar from other areas are conspicuously absent in this vault, although the very rapid retrieval methods imposed on the excavation of the final stage of the project may have influenced the recognition or otherwise of very small deposits of secondary refuse. Nevertheless, only 2 of the 39 interments can actually be demonstrated to have been moved after their initial deposition. The chaos apparent in the upper deposits in this vault can be inferred to have been caused by a combination of abandonment deposition and graviturbation, ie upright interments falling over and damaging other material.

The primary sequence appears to have occurred as follows. Eight slabs of roughly cut limestone of various sizes had been placed on the floor, towards the south-east corner; these were employed as coffin supports. Interments were initially placed in a single stack against the east wall of the vault, up to a height of five; none of these were dated. An upright interment of 1747 was placed in the north-west corner, and four others, undated, were placed to the west of the first stack, in a row aligned east–west or west–east. North of these an interment of 1749 (2939) was placed on the same alignment. Another upright, undated, placed next to 2918 in the north-west corner, may be of the same period.

Two interments, one of them dated to 1768 (2963), were placed in the tunnel in the south wall of the vault. An upright coffin (2927) was placed in the south-west corner of the vault in or after 1773, and another of 1773 was placed adjacent to it, also upright. From the same year an interment was deposited on the floor of the vault immediately east of the brick plinth supporting the steps. This would have had implications for later access to the area, and it is entirely possible that its final position should be dated much later than its original location.

The first coffin in a stack of four in the extreme north-east corner was a secondary interment dating from 1791. Two of the others in the same stack were also secondary, and it is possible that this stack was created in one period of activity after 1847, when the latest of them was introduced. Five further interments were added to those on the floor in the centre of the vault, and then three more were added facing north–south or south–north followed by a fourth layer, on the same alignment, and another single coffin to the north.

An infant interment (2947) dated to 1734 was deposited as a secondary interment on the floor to the west of these stacks, perhaps in the 1790s, next to an upright (2929) which had no date. The remaining floor space was occupied by a north–south interment (2945). One more was added to this, and two more to the interment just below the steps. Interment seems to have been much more frequent during the second half of the 18th century than in the first half of the 19th. Of the body or 'stone sarcophagus' of Edward Peck, Commissioner and owner of the vault, who died in 1736 and is commemorated in a memorial in the sanctuary of the church, no evidence was forthcoming. The abandonment of this area after 1847 probably coincided with the 1866–7 abandonment phase. Inscribed slabs relating to the Peck family (Appendix C11, C13, and C17 and Figs 3.30–2) were smashed and included in a sealing deposit of several tonnes. During this deposition two of the upright coffins seem to have collapsed and fallen over, and consequently appeared almost to have been thrown in with the dumped material.

8 The parochial area (PV)

This area contained 335 interments:

Adults	252/335/75.2
Infants	083/335/24.8
Primary	292/335/87.2
Secondary	043/335/12.8
Dated	119/335/35.5

The parochial area covers approximately 50 m^2. At the west end it includes the lower central east tunnel, extending to the east end of the Lemaistre/Pontardant vault. This tunnel emerges at the west end of the apse which forms part of the tower foundation and is raised above the remainder of the vault on a brick plinth approximately 0.2 m above the rest of the floor. The remainder forms a roughly square area corresponding to one and a half bays of the nave of the church. Two piers linked by an arch effectively divide the square into two, with a north and south transept immediately to the east of the apse.

Figure 3.29 Interior of PK looking west (PK:0000.01)

Figure 3.30 Memorial to John Peck recovered from PK

Figure 3.31 Memorial to Edward Peck recovered from PK

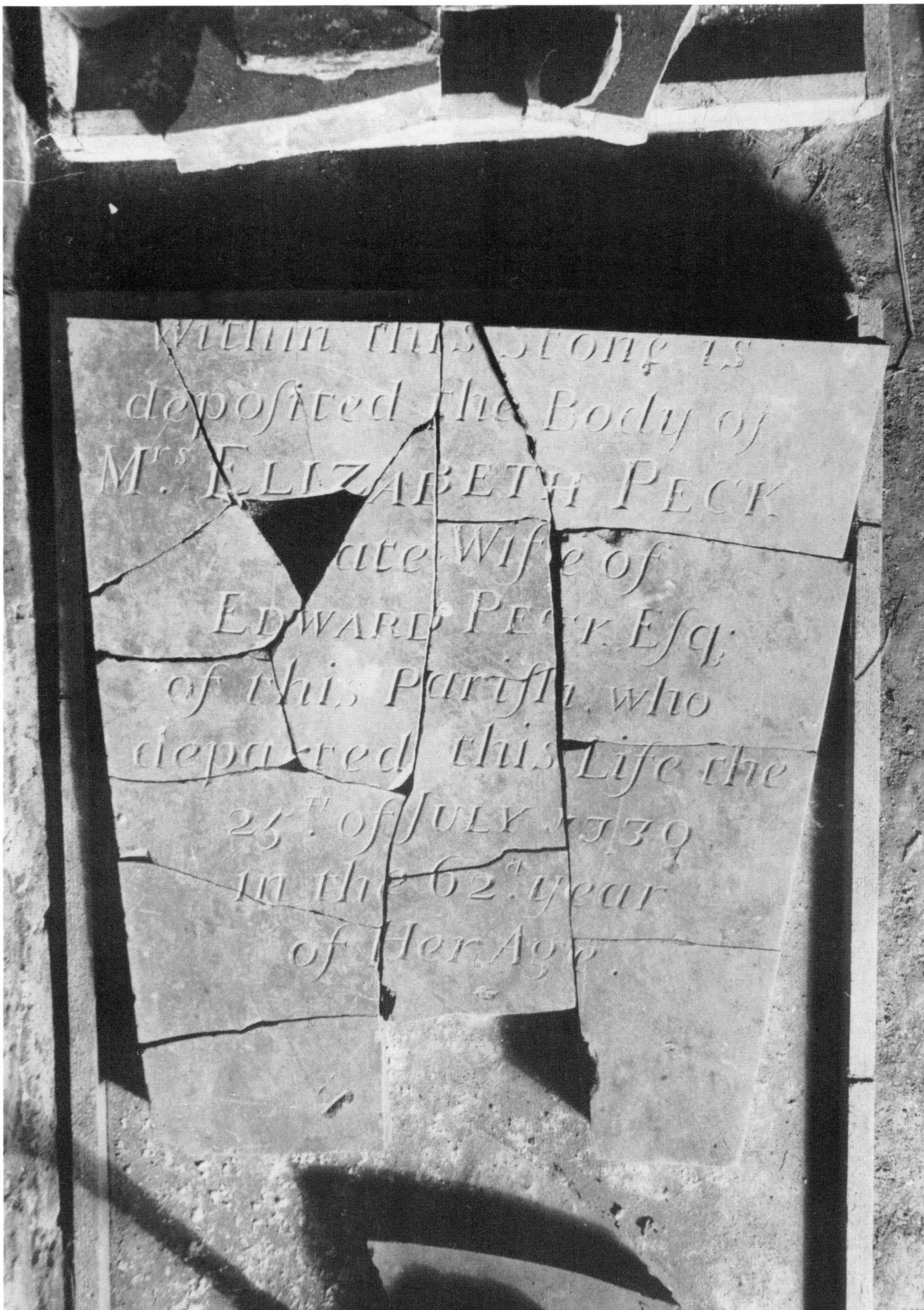

Figure 3.32 Memorial to Elizabeth Peck recovered from PK

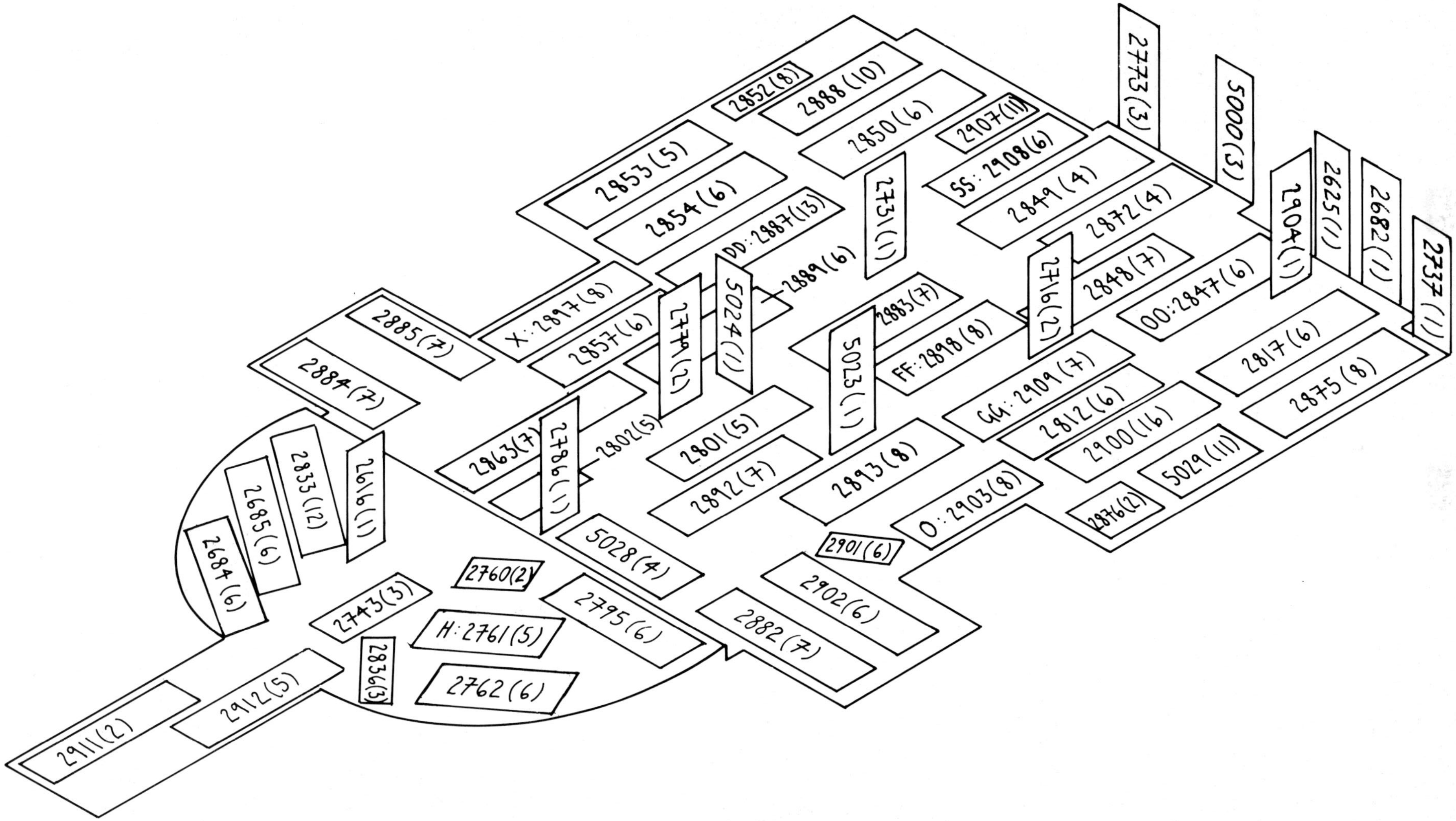

Figure 3.33 The parochial area: burial scheme

Figure 3.34 'Ghost' of passageway at east end of PV

At the east end a blocking wall was created between two piers, which separates the main parochial area from the eastern parochial area (EP) and originally separated it from the rest of the crypt. An entrance into the vault was provided at the east end by a doorway built into the blocking wall. However, as noted in section 3 above, access was also possible from the west end during certain periods of the use of the vault, and at the time of the abandonment phase was the only means of introducing the sealing deposits into three areas, including the PV and EP areas (see Fig 1.5).

That this area was a public vault there is little doubt. An inscription located on the east side of the east wall of the EP area records the closing of the 'Public vault' (Appendix C26). Furthermore, the number of interments located in this area precludes its identification with a single family. Indeed, a minimum of 87 families are represented in more than 330 interments.

The interment activities here follow a distinct, although somewhat idiosyncratic pattern, over nearly 80 years, between 1741 or perhaps earlier, and 1813/16. The ambiguity over the date of the last interment is a problem discussed below. Figure 3.33 shows burial deposits and identifies stacks where they are specifically mentioned in the text.

Initially interments were deposited in the apse and the tunnel at the west end. The earliest date is provided by a secondary interment (2769) of 1741, located low down in stack H in the southern half of the apse. Dates are scarce from this part of the vault, and it is entirely possible that interments were being deposited here from the early 1730s onwards. However, the role of the area beneath the church steps which was discussed above with the LP area may suggest otherwise, if it was used as a public vault during that period.

By the beginning of the 1760s interments had been stacked up to four high over the whole area of the apse, and were also being deposited in the two transepts and against the north and south walls of the vault. Perhaps because of the awkward shape of the apse, coffins do not seem to have been stacked to maximize the use of space. Indeed, throughout the period of the use of this area deposition seems to have taken place in an apparently haphazard way, in striking contrast to the 'neat and tidy' regime displayed by the pattern of interment in the eastern parochial area (EP). However, it is significant that

Figure 3.35 General view of the PV during excavation (PV:0528.01)

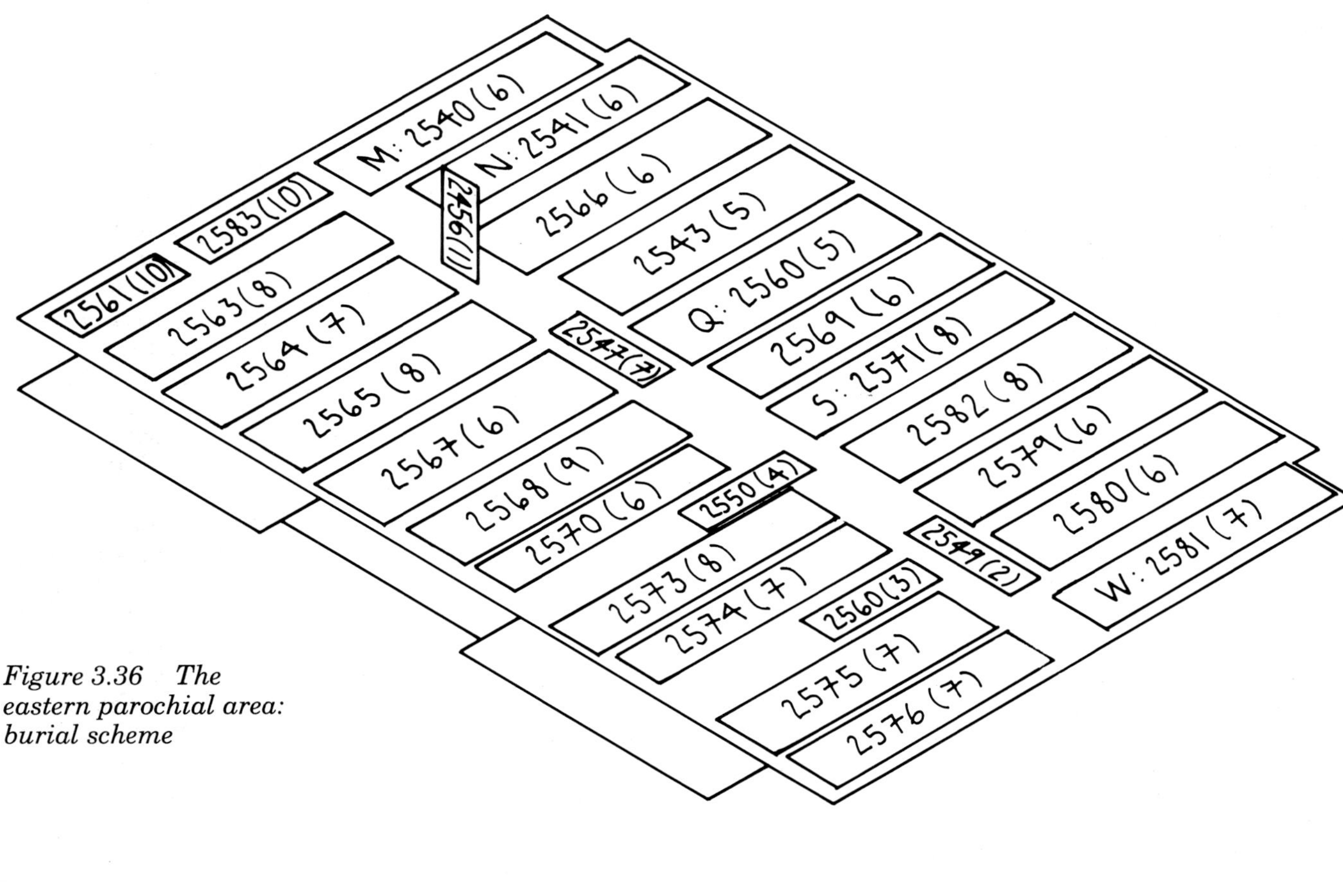

Figure 3.36 The eastern parochial area: burial scheme

the density of burial was almost identical in the two vaults, at around 1/m^3. After a period during the 1760s and 1770s when coffins were being stacked two and three high around the edges of the vault, interments began to cover the floor of the entire area inwards and east towards the entrance, apparently rarely more than one coffin high. By the late 1770s and early 1780s the apse was virtually full, with interments stacked up to six high in places and coffins then being manoeuvred into remaining gaps between earlier interments. Until the beginning of the 1790s space still does not seem to have been regarded as a critical resource. However, during the period 1793–5 large-scale maintenance procedures appear to have been instigated.

These procedures have been identified by unconformities in the dated sequence within the vault. It is notable that the quantity of secondary refuse (coffin debris) created elsewhere in the crypt during maintenance and rearrangement was largely absent from the parochial area, and that the rearrangement of the vault in the 1790s could hardly have been identified without a high proportion of dated contexts. It is very likely that a higher proportion of dated interments would have led to the identification of more maintenance activities than those rather crudely reconstructed here. It is inferred that the refuse created by this activity was removed from the vault entirely, and burnt or disposed of elsewhere.

The large-scale movement of coffins in the parochial area probably began in 1793. The floor of the vault must have been almost entirely covered with primary interments, and in consequence it would have been difficult to introduce further deposits without trampling on and breaking those already lying on the floor. Interments were removed from an area roughly bounded by stacks OO and SS at the east end, and GG and DD in the centre (see Fig 3.33). The stacks previously created from O across to X were the repositories for many of these displaced coffins, while others were stacked to the north and south of the cleared area. A space was left which formed a passageway from the entrance to roughly level with the two outstanding piers (the 'ghost' of this passageway can clearly be seen in Fig 3.34). Other coffins were placed upright around the edges of this passage (see Fig 3.33), including one (5000) which was fixed to the east wall immediately south of the entrance by means of an iron bracket.

After this rearrangement it would have been possible to reach most of the stacks around the edges of the vault from the passage. However, the next primary interments seem to have been deposited in the passage, leaving an access-way less than 1 m across. This reinforces the inference that control of space as a resource was not the forte of the sexton during this period.

By 1805–6 most of the vault was stacked six high with coffins, and only those stacks at the east end and the space immediately to the west of the entrance were accessible. It is likely that intermittent periods of maintenance and rearrangement were taking place up until 1810 or thereabouts, although it is difficult to identify the extent or nature of these activities.

By July 1812 the vaults were so full (as is obvious from Fig 3.35) that the Church Council sent a committee, including an apothecary, into the vaults to assess their state. The result was a resolution of July 1813 to close and seal the vault in order to prevent what was seen as a grave public health risk. The vault was duly closed and the second public vault (EP) opened in the same year. It was also at this time that it was determined that no further burials should take place in the crypt unless they were enclosed in lead caskets.

One further activity took place in the vault prior to the abandonment phase of 1866–7. An interment of 1816 (2860) was introduced from the west end of the vault, almost certainly through the dismantled wall at the east end of the upper central tunnel (UC). It was taken to the east end of the vault. Five coffins were removed from stack FF. The new interment was placed in the hole thus created, and the coffins were then replaced above it. The entrance from the west was then resealed.

Any number of macabre possibilities have been suggested to explain what appears to have been an extraordinary combination of conspicuous effort and covert motive. However, beyond the obvious fact that this interment was illegally made without a lead casket (as were many other more public interments after 1813), it seems idle to speculate on the reasons for such a specific activity.

9 The eastern parochial area (EP)

This area contained 181 interments:

Adults	128/181/70.7
Infants	053/181/29.3
Primary	099/181/54.7
Secondary	082/181/45.3
Dated	150/181/83.3

The eastern parochial area, or second public vault, was opened in the summer of 1813 immediately after the closure of the first public vault (PV) (Appendix C21). In area it is considerably smaller than the PV area, covering some 28m^2. The density of burial is very similar, despite the apparently more organized and efficient way in which the EP area was filled. It will be apparent from Figure 3.36 that the size and shape of the EP area were convenient for high-density interment, and this was exploited: at 4 m x 6 m the vault is roughly the same size as the ground floor of a typical 'two up two down' British terraced house, and contains more than 180 coffins (see also Figs 3.37, 3.38).

Interment took place here between 1813 and 1845, when the vault was closed. With more than 80% of the interments dated it has been possible to identify with some confidence the proportions of primary and secondary interments here. It is quite likely that an area such as PV contains many more secondary interments than have actually been identified; a more

Figure 3.37 General view of EP from its west entrance (EP:0000.03)

Figure 3.38 General view of EP during excavation (EP:0700.01)

accurate percentage may be that derived from the EP area.

The earliest interments were probably placed against the west wall of the vault, but periodically moved onto stacks in the north-west corner. Such rearrangements seem to have been frequent and on a small scale, and little secondary refuse was apparently created, or remains, from such activity. Many of the coffins were in good condition even at the time of excavation, and the stacking seems to have been 'neat and tidy'. Infant coffins were kept together in two stacks in the extreme north-west corner, to an eventual ten high.

Primary sequences occurred in the two of the other corners during the mid-1820s. Two stacks were created between 1823 and 1827 at M and N, and another between 1819 and 1824 at W (see Fig 3.36). In the south-west corner primary burials of the early and mid-1820s were overlain by secondary deposits. From this period onwards a similar pattern of single primary stacks and frequent rearrangements continued until the end of the 1830s, by which time there were interments on the floor immediately in front of the entrance at S and Q (see Fig 3.36). The lateral north–south space left between the west row and the east row was filled, probably after this time and continually until closure of the vault, with infant interments, mostly apparently squeezed vertically in between the heads and ends of adult coffins. Between 1839 and 1845 primary interment continued, but seems to have been interspersed with the introduction of a greater number of secondary interments, probably from elsewhere in the crypt, which occur at the top of many of the stacks in this area. There is a striking contrast between secondary interments in this vault and in the south and north chasms and the north parochial area. In the EP 89.3% of coffins had lead shells; in the south chasm (SC) 7.0% had lead shells; in the north chasm (NC) and northern parochial combined, there were none. The implications of this are variously discussed in chapters 4 and 5.

4 Christ Church as mortuary site

Here lie I at the Chancel door
Here lie I because I am poor
The further in the more you'll pay
Here lie I as poor as they.
(From an inscription in Kingsbridge church, Devon, 1795, quoted by Boore 1985, 23)

1 Sources of evidence

Post-medieval mortuary practice has only recently become a subject of archaeological investigation, but the previous lack of archaeological work in this field testifies to the neglect of this important branch of mortuary studies. What little archaeological investigation which has taken place in post-medieval contexts has been constrained by a lack of resources. Burial archaeologists everywhere are conscious of the difficulty of reconstructing the liturgy and eschatology of past societies from their mortuary remains. At Christ Church it was hoped that a combination of a high level of archaeological interpretation and comprehensive historical source material would demonstrate the potential for a more confident reconstruction of the death ways of the period 1729–1867. Somewhat surprisingly, the mechanics of death in this period are not as easy to understand from strictly historical material as was first thought.

Historical sources for the period up until the beginning of the 19th century are scarce compared to the huge volume of material on death which was generated during the Victorian period. Mortuary practice prior to the onset of the Industrial Revolution has been the subject of Gittings' excellent *Death, burial and the individual in early modern England* (1984). For the latter two-thirds of the 19th century Morley's *Death, heaven and the Victorians* (1971) provides a useful synthesis of literary and documentary evidence. For the Victorian period the most important primary sources are the London directories, starting with *Johnstone's Commercial Directory* of 1817, and a series of reports to Parliament on the state of London cemeteries. These include Edwin Chadwick's *A supplementary report on the results of a special inquiry into the practice of interment in towns* (1843) and *A report of a general scheme for extramural sepulture*, produced as a blue paper for Parliament in 1850 (Lewis 1850).

For the period 1729–1817 the sources are sparse, in the main because undertakers' accounts rarely survive, and because directories provide insufficient detail. Trade cards such as those reproduced in Figures 4.1–4.5 are a useful supplement to what little we know. *The Times*, which began publication in 1785, records many incidents concerning funerals and crimes involving theft of bodies or coffins and malpractices in the undertaking trade. For the social context for the post-medieval funeral in London, Charles Dickens is especially illuminating in such works as *Martin Chuzzlewit* and *Bleak House*. The seminal work on London in the 18th century is Dorothy George's *London Life in the eighteenth century* (1925). For Spitalfields itself the Christ Church Burial Register survives, along with the invaluable Hurlin letter (Appendix B1). It is extremely fortunate that one of the rectors interviewed by Chadwick was the Reverend William Stone, incumbent of Christ Church during the 1840s. A list of relevant legislation and other documents appears in Appendix A. The interested reader is advised to use the bibliographies contained in those secondary works mentioned above.

2 The burial vault as a resource

The post-medieval burial context may be seen as the result of a series of opportunistic responses to ideological and pragmatic needs, both real and imagined. The identification of the burial vault as a resource stems from its availablity as a space for the disposal of the dead and its potential as a generator of revenue for the church through sale of that space. At Christ Church, as at any other parish church, those who died in the parish had a right to be buried in the graveyard. Depending on their financial standing they might choose a more or less expensive funeral, and they might choose to be laid to rest not in the churchyard, but within the walls of the church itself. In order to understand the importance of that choice for individuals who died in the period from 1729 to 1867, three perspectives must be considered: that of the individual, that of the church, and that of the undertaking industry, which grew to control the business of death during the 18th and 19th centuries.

In both spiritual and temporal terms the disposal of the Christian dead had certain necessary requirements. In spiritual terms this involved the necessary repose of the soul in readiness for the day of judgement. In temporal terms it required, prior to the reintroduction of cremation at the end of the 19th century, the procurement of a space in which to be interred. Attitudes during the 18th and 19th centuries towards disposal are a complex expression of the ways in which these resources were controlled.

The desire to be placed in death as close to a source of spiritual power as possible has been a feature of Christian belief for more than a thousand years, and finds expression in the placing of burials close to early Christian shrines such as those at Church Island, County Kerry, Eire (Thomas 1971, 69) and Ardwall Island, Kirkcudbrightshire, Scotland (*ibid* 72). The more important the shrine, the more desirable the access to it, as evidenced from the medieval period up to the 20th century by the occurrence of high-status burials of secular and religious figures in edifices such as Westminster Abbey and St Paul's Cathedral, although in these instances it might be a national or secular motive, as opposed to a religious one, which attracted high-status burials.

It might be expected, therefore, that in 18th century London prestigious places of worship such as Christ Church would attract wealthy people in death as in life. This was indeed the case. However, it is difficult to find evidence that nearness to a holy place was a primary consideration in choosing the place of disposal, even though there is abundant evidence that burial vaults beneath churches were highly desirable places to be buried.

Certainly there were strong personal reasons for being interred in a vault. William Hurlin, writing of his grandmother, Sarah Hurlin, who was interred in the crypt at Christ Church in 1839 (Appendix B1), was of the opinion that 'she shrank from the idea of being buried in the earth.' The notion expressed here that earth burial is 'dirty', and that there was a repugnance towards it due to the knowledge that the flesh of the corpse would be consumed by worms and other small fauna, is easy for 20th century individuals to comprehend, for it is a common enough sentiment today. The present alternative, which the people of Christ Church would have found morally outrageous (Morley 1971, 91), is cremation. Before cremation became an acceptable rite an alternative to earth burial was interment within a burial vault. This must have seemed cleaner, tidier, and somehow more hygienic in spite of the fact that bodies interred in crypts are consumed, if more slowly, in similar ways to those in churchyards. The belief illustrated by the inscription at the start of this chapter, that it was warmer or cosier inside, may have been widespread.

Security of tenure seems to have enjoyed a high priority for the deceased during the 18th and 19th centuries. Burial in a churchyard posed two risks. Firstly, the actual tenure guaranteed by a church was limited, literally, by its turnover. As population increased, the number of dead needing disposal increased (George 1925, 111) and churchyards became notoriously overcrowded. Not until the opening of the great metropolitan cemeteries in the 1830s did the situation begin to ease in urban churchyards. Even in 1853 Charles Dickens was able to evoke the grim atmosphere of a London churchyard:

> O, what a scene of horror! 'There!', says
> Jo, pointing. 'Over yinder.
> Among them piles of bones, and close to
> that there kitchen winder! They put him
> wery nigh the top. They was obliged to
> stamp upon it to git it in. I could unkiver
> it for you with my broom, if the gate was
> open. That's why they locks it, I s'pose',
> giving it a shake.
> 'It's always locked. Look at the rat!', cries
> Jo, excited. 'Hi, look!
> There he goes! Ho! Into the ground!'
>
> (*Bleak House*, 1853)

The second risk was posed by the 'resurrection men', or 'sack-em-up gentlemen' who operated in the capital. During the height of their activity, between 1750 and 1830, these operators supplied hospitals such as Guy's and St Bartholemew's, as well as small, private anatomy schools, with bodies for dissection and study (Hewer 1954, 154). The 'going rate' for a body during much of this period was about four guineas (Macphail 1914, 113), and hundreds of people seem to have made a handsome living from it. The practice was the cause of considerable concern to the general public. *The Times* often reported such incidents. On 1 January 1787 it recorded that four body snatchers had taken a corpse from the churchyard of St Saviour's, Southwark, and brought it away in a coach. A reward of five guineas was offered for the return of the corpse. Not until 1832, when the 'Anatomy Act' (2 & 3 William IV c.75) was passed, which made provision for the medical colleges to use the unclaimed corpses from the poor house for their study, did body snatching fall into a rapid decline.

Evidently the security provided by the walls of a crypt was an attractive alternative to the thought of being disturbed in this way. Furthermore, the tenure for a space in a vault was generally very much longer than in a graveyard. Thomas Jervis, a silk thrower and trustee for the parish almshouse, witnessed the purchase of a vault by Philip Dutch, a relative of his. The deed, which still survives, records the right to bury in one of the unexcavated vaults at Christ Church in January 1735 'in perpetuity'.

If interment in a burial vault was desirable, it was also expensive. The very expense may itself have provided an incentive for burial there: exclusivity. Indeed, the entire process and paraphernalia of death 'served to maintain the *status quo* and to reaffirm the traditional hierarchy of power and prestige' (Gittings 1984, 89).

Whilst the desire to be buried in a crypt might have found expression in social attitudes, or in the specific instructions of a will, a still more powerful factor controlling the circumstance of disposal may have been the notion of correctness, of conforming to what was 'proper': 'Indeed, it is doubtful whether many of the bereaved actually saw themselves as making conscious choices concerning funerals, but were simply carrying out what was customarily agreed to be appropriate. However, these customs were gradually modified by the users, possibly quite unthinkingly at times, in response to changing attitudes and circumstances' (Gittings 1984, 86).

The 'changing attitudes and circumstances' during the 18th and 19th centuries encompass many social phenomena: the rise of individualism, the Industrial Revolution, and the reactionary reformism of the Victorian Anglican church. The role of the church in this period was crucial in controlling the resources required by the dead. If the most important of these resources were space and security, then the church, until the 1830s, held a virtual monopoly. It is noted in chapter 1 that the Act which set up the Commissions for Fifty New Churches in 1710–11 expressly forbade burial in the vaults of those churches. The minutes of the Commission, however, (see Appendix A) demonstrate that even prior to the consecration of those churches burial in their vaults had been anticipated by the church councils. Their motive seems to have been a financial one. The Commissioners' churches, and many others, suffered a cash crisis from the moment of planning onwards. The cost of construction and upkeep was enormous, and there was a duty, as well as a public demand, for burial. The 68,000 burials registered at Christ Church represent approximately 490 burials per year, in both the crypt and churchyard. The income from these averaged £215 per annum during the first four years of the incumbency of the Reverend William Stone (Chadwick 1843, 100). The 1000 or so burials in the vaults at Christ Church must alone have raised in excess of £2500. In the 'Table of Fees' of July 1729 (Greater London Record Office, MS) the church warden was to be paid £1 for every burial 'in any vault under the church'. He was to receive £10 'for every vault of nine feet in length and seven in breadth'. This compares with a fee of 11s for a plot in the most expensive, western part of the graveyard. Philip Dutch, mentioned above, paid £35 8s for his family vault at Christ Church.

An important role in the finances of the church was played by the warden, who had the right to levy a parish rate for the poor, and for repairs to the church. He needed an independent source of income apart from the church because he was obliged to settle any outstanding accounts from his own finances if no parish money was available. It would appear, therefore, that for the church warden at least, burials, and especially burials in the crypt, were a valuable source of income. Other individuals benefited. The rector received a substantial share of the burial fees towards his living — £125 per annum according to Chadwick (1843, 100). The clerk of the parish received 2s for every burial in the vaults, while the sexton received 1s. The rector himself concluded in 1843 that 'under the present system, the burial of the dead is a source of profit; it yields an annual surplus towards defraying the other expenses of the church' (Chadwick 1843, 100). At Christ Church and elsewhere an additional source of profit came from burials which were conducted for people from outside the parish. The 'Table of Fees' states that 'all strangers except those inhabiting the artillery ground and the place of Norton Folgate pay double all & every these fees and also all parishioners that go to other churches to be buried are to pay the usual fees before mentioned.' Given that individuals interred in the crypt at Christ Church came from more than 60 parishes (M Cox, pers comm), we may infer that for the church burial was a lucrative business.

It appears that sextons and their employees were not above making extra profit. Morley (1971, 37) cites evidence that the job of the sexton or gravedigger was such an unpleasant one in Victorian England that the men were given to bolstering their morale with considerable quantities of alcohol. It was not uncommon, according to one gravedigger who came before the 1842 Select Committee on Improvement of Health in Towns, for gravediggers to steal the lead from coffins, to sell the bodies to surgeons, and burn the coffin wood for warmth. It was even alleged by an undertaker that Dissenting ministers were known to speculate on burial grounds. He said 'it has been remarked to me that they get more money by the dead than by the living', a quip on their official income (quoted by Morley 1971, 38).

3 The funeral industry in London 1729–1867

The Church may have propped up its finances by exploiting its control over burial space, and individuals may have profited more or less illicitly from that control. However, compared to the funeral industry which grew alongside during the Industrial Revolution in the 18th and 19th centuries, the Church was a rank amateur. People who undertook to arrange funerals saw an immense source of income in laying on a 'traditional' funeral, themselves dictating what was traditional and what was proper.

The trade of funeral undertaker developed in London towards the end of the 17th century (Gittings 1984, 94). An undertaker initially seems to have been one or other of the tradesmen who supplied services or furnishings for a funeral, and more especially for a middle-class or aristocratic funeral. The tradesman would operate as a contractor, undertaking to arrange the various components of the funeral around his own trade. Thus the first known undertaker, William Russell, was a coffin maker and painter who came to an arrangement with the College of Arms in 1689. The College agreed to attend funerals organized by Russell for a fee (Gittings 1984, 96). So rapid was the development of this industry that the aristocratic and heraldic origins on which the middle-class funeral was founded were quickly forgotten. The functions initially performed by members of the College of Arms were rapidly usurped by the undertaker and his employees. By the middle of the 18th century undertaking was an established profession in the capital, encompassing not only the immediate needs of the bereaved in organizing a funeral, but appraising and selling the contents of houses, arranging the service with the church, and advising on the type of funeral which was seen to be appropriate for the social circumstances of the deceased. Puckle (1926, 98) cited a case concerning the funeral organized by a family in a respectable street in a London suburb. The family

Figure 4.1 Trade card: Wm Grinly, coffin maker, 1745

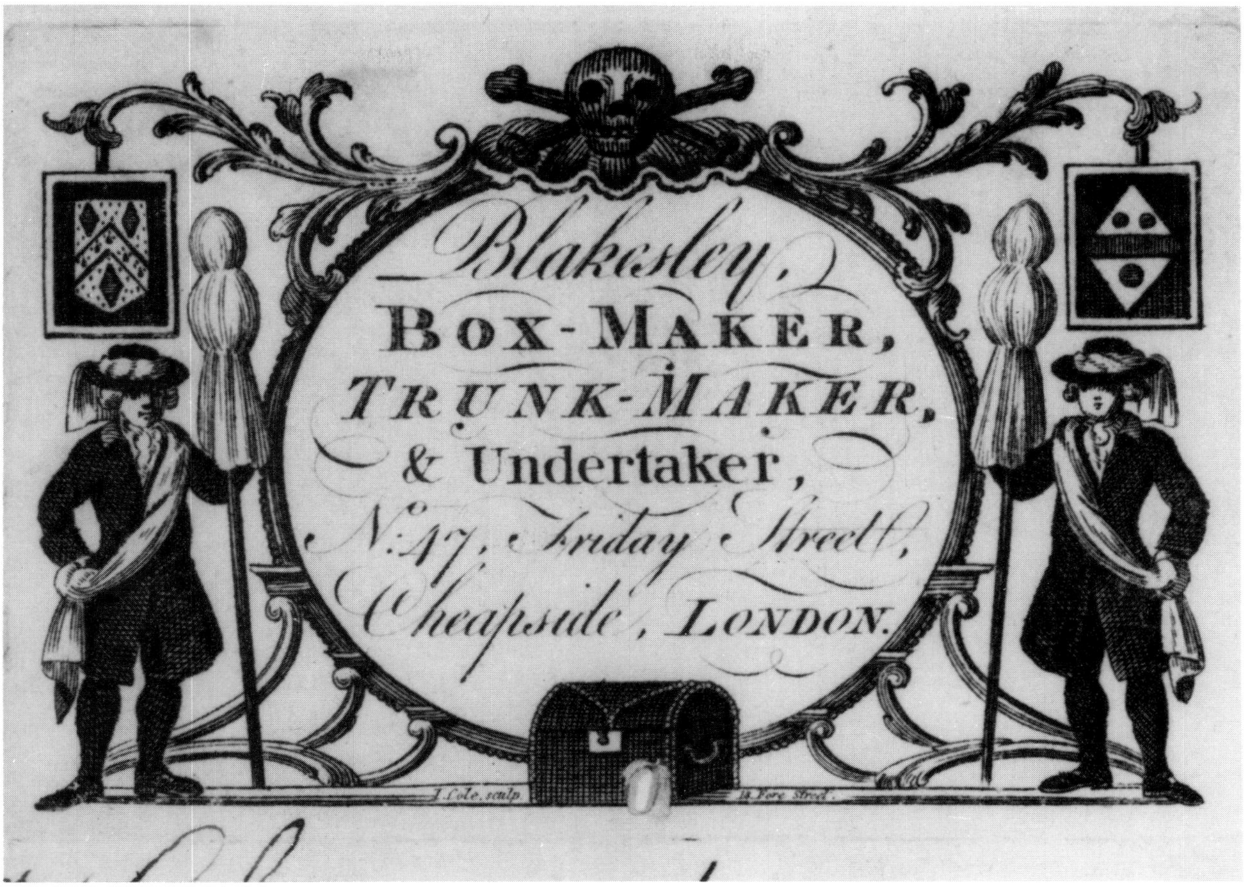

Figure 4.2 Trade card: Blakesley, box maker, trunk maker, and undertaker, 1792

of the deceased were apparently determined to forego the usual pomp and asked the undertaker to provide a plain elm coffin without ornament. The undertaker professed himself horrified, declaring that only polished oak was suitable for the district. The profession came to be regarded with contempt:

> The last trade I shall mention is the Undertaker, a set of men who live by death, and never care to appear but at the End of a Man's Life, they may then properly enough serve to bring up the Rear of our Trades; their Business is to watch Death, and to furnish out the Funeral Solemnity, with as much Pomp and feigned Sorrow as the Heirs or Successors of the Deceased choose to purchase: They are a hard-hearted generation, and require more money than Brains to conduct their Business; I know no one Qualification peculiarly necessary to them, except that is a steady, demure and melancholy Countenance at Command: I do not know that they take Apprentices in their Capacity as Undertakers, for they are generally Carpenters, or Herald-Painters besides; and they only employ as Journeymen, a set of Men whom they have picked up, possessed of a sober countenance, and a solemn melancholy Face, whom they pay at so much a Jobb.
>
> (Cambell 1747)

In a case cited by Morley (1971, 38) an undertaker went to the extreme of challenging the monopoly of the church on burial:

> A 'burial ground' in the parish of St John's, Southwark, belonged not to the church, but to an undertaker: he had bought a cellar or 'vault' that ran under four houses, and filled it with wooden (not lead, as was customary in 'vaults') coffins; the undertaker's man, who was, of course, not ordained, regularly donned a surplice and read the service 'profanely'.

Until the beginning of the 19th century undertakers were generally people who operated a number of trades themselves and had access to many more. The purchase ledger of Richard Carpenter, a

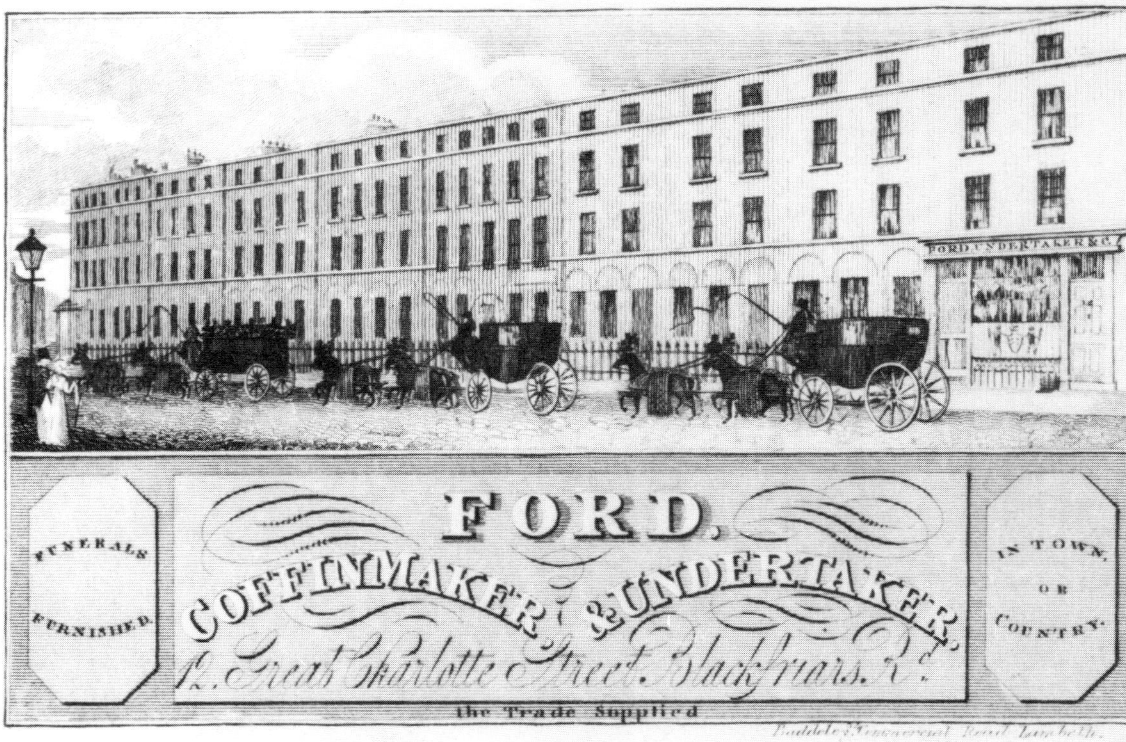

Figure 4.3 Trade card: Ford, coffin maker and undertaker, 1825

Figure 4.4 Trade card: Joseph Turner, coffin maker and furnishing undertaker, 1838

London undertaker, for the year 1746–7 illustrates the diversity of trades which an undertaker needed to produce a funeral. Carpenter had 28 partnerships and traders in his ledger. There were plumbers, a founder producing coffin nails, a man who hired palls and hoods, another who hired feathers, others who provided gloves, lace and crepe, and an engraver. A Mrs Ann Williams made coffins and a Mr D Mesman supplied *rich allamode*. Figures 4.1 and 4.2 illustrate this diversity. By the end of the first quarter of the 19th century the industry had become more specialized and highly competitive, as a trade card of 1825 shows (Fig 4.3). During the early Victorian period the number of undertakers was thought to be excessive. In 1843 the Post Office Directory for London listed 275 undertakers, 258 undertakers and cabinet makers, 51 undertakers and builders, 25 undertakers and appraisers, 19 undertakers and auctioneers, 7 undertakers and house-agents, 3 undertakers and fancy cabinet makers, and 2 undertakers and packing-case makers. The 1850 Commission (Lewis 1850, 34) was told that a dozen would suffice. So fierce was rivalry that elaborate advertising was commonplace, as shown in Figure 4.4, and undertakers such as Mr Saddler of Cripplegate were performing funerals up to 70 or 80 miles away and offering credit over periods of up to 25 years (Lewis 1850, 36) in order, it seems, to attract custom.

There is no reason to believe that the trade of the undertaker was carried on in Spitalfields differently from anywhere else in London. The first direct evidence that an undertaker was operating in Spitalfields comes from *Johnstone's London Commercial Directory* of 1817, in which M Felton of 47 Brick Lane was listed as Upholder and Undertaker. By 1852 the *Watkins London Directory* listed six undertakers for Spitalfields. There are earlier signs that the industry was being conducted in the parish, especially in the records of the Worshipful Company of Upholders, which incorporated undertakers. The list of Freedom Admissions for the company between 1771 and 1798 includes five upholders resident in Spitalfields. During the same period there are a hosier, a coach-builder, two tin-plate workers, several mercers, and many weavers. It may be inferred that the requisites for performing a funeral in Spitalfields were readily available well before 1817. Certainly a market for such skills existed. Not only residents of Spitalfields, but those of Shoreditch and Bethnal Green nearby, and others from all over Lon-

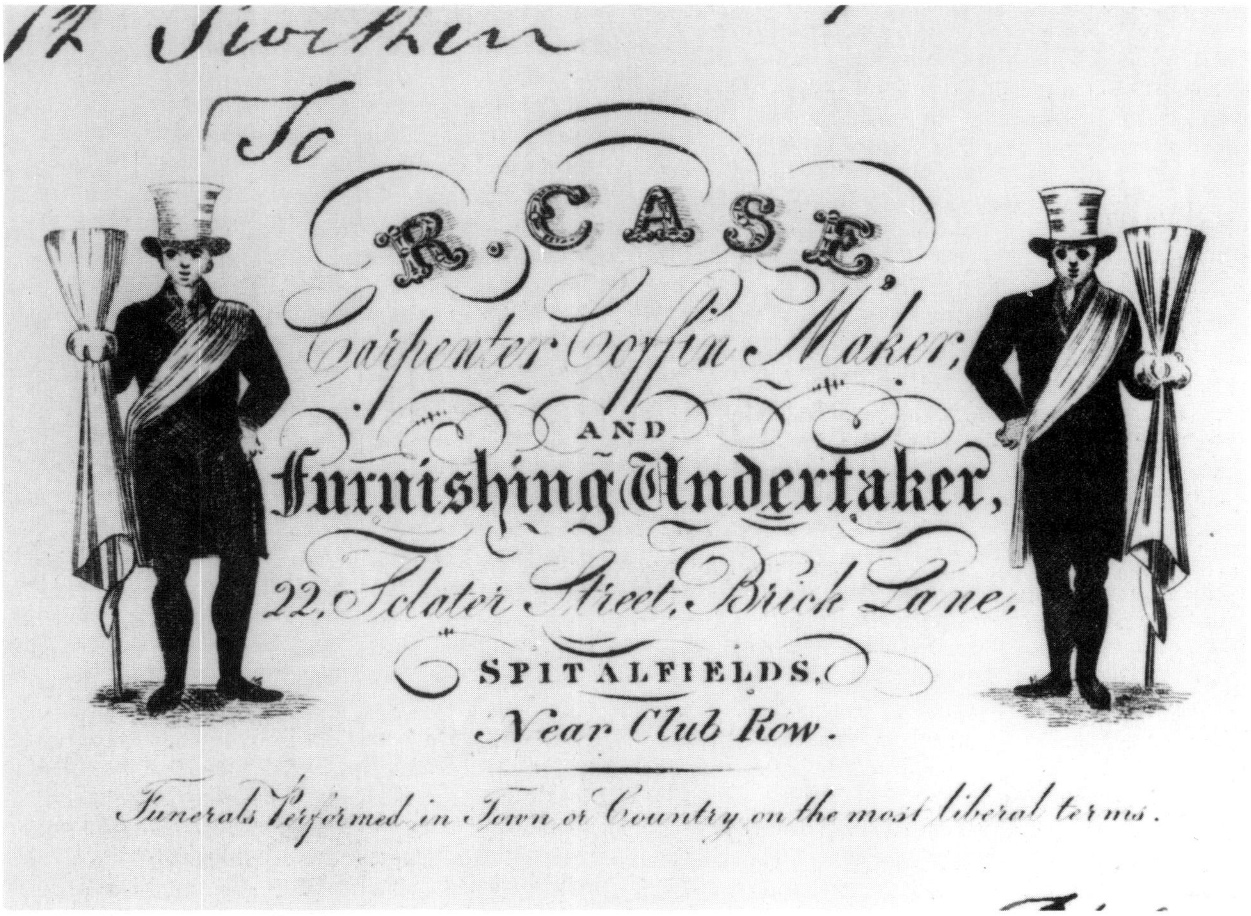

Figure 4.5 Trade card: R Case, Carpenter, coffin maker, and furnishing undertaker, 1829

don, wished to be buried in Spitalfields. As late as the 1840s the Reverend William Stone of Christ Church complained to Edwin Chadwick that undertakers were smuggling in bodies from outside the parish in order to avoid paying the double fees which were charged in Spitalfields, as they were elsewhere in London (Chadwick 1843, 46), for the burial of non-parishioners.

Those who undertook to arrange funerals, especially during the early and middle 18th century, came from diverse trades, most of which would have occurred locally in many areas of London, and perhaps especially Spitalfields with its background in the cloth trade. Carpenters and milliners might call themselves undertakers in order to attract such custom. It is likely that they would cooperate to provide the services for a funeral, contracting the construction of lead coffins to a plumber, and the provision of coffin ornaments to a tin-plate worker. By 1817 a degree of specialization had taken place (see Fig 4.5) and all the resources required for the arrangement of a funeral were well established within the parish. Productivity was high in the undertaking trade by the beginning of the 19th century, and grew steadily. Mr Saddler, of Cripplegate, claimed in 1850 that he had been burying 1800 people every year for the previous 32 years. The high mortality rates of this period provided an income for thousands of people in the capital. Chadwick (1843, 53) thought that there were 'very few wealthy undertakers. They are described by one of them "as being some few of them very respectable, but the great majority as men mostly in a small grubbing way of business." ' Dix, an undertaker interviewed by Chadwick, said:

> I frequently perform funerals three deep: that is, I do it for one person, who does it for another who does it for the relatives of the deceased, he being the first person applied to. Everybody calls himself an undertaker. The numerous men employed as bearers become undertakers, although they have never done anything until they have got the job. I have known one of these men get a new suit of clothes out of the funeral of one decent mechanic.
> (Chadwick 1843, 51)

If most of those in the business were lowly, there were others who made a more substantial trade. J Bedford of King Street, who owned a coffin furniture warehouse, claimed in 1850 that 90% of the London trade in coffin furniture was done by his firm. Another firm, Harrod's, expanded to include a wide range of furnishings and became the celebrated Harrods of Knightsbridge.

The actual provision which was thought necessary for a funeral for a 'decent' person does not seem to have altered greatly during the period 1729–1867. The bill for the funeral of Mrs Mary Hazmer, buried on 21 April 1764 by Richard Carpenter, undertaker, lists most of the items likely to have been needed. The bill was jotted down in pencil, and not all of it is still legible. Items are costed in pounds, shillings, and pence.

Elm [coffin] cover[ed] with cloth drape, double lead plate, 4 pairs of large chinia[?] hand[les]	6.0.0
shroud sheet and pillows	1.18.0
4 men to move body from Kentish Town	0.8.0
lead plate & writing for the side of the coffin	0.2.0
To a hearse & pair	0.11.0
paid turnpike	0.0.3
4 funeral tickets delivering	0.4.0
use of best pall	0.7.6
lid of feathers	1.0.0
5 gentleman's cloaks	0.5.0
5 new crape hatbands	0.17.6
1 silk hatband [for the] minister	0.7.6
cash gave [to] the clerks by order	0.5.0
4 scarfs & hoods	0.6.0
5 [pairs of] men's lace kyd [gloves]	1.0.0
5 warm ja[ckets]	1.5.0
2 [mutes] at [the] door with staves capes and covers	0.10.0
2 hats [and] 2 pair[s of] gloves	0.6.0
Hearse & 4 horses to St Dunstans	0.18.0
feathers	1.15.0
velvets	1.0.0
6 pages to attend hearse with caps & Truncheons, black with favours and gloves	1.6.0
2 mourning coaches	0.18.6
4 pages to attend the coaches & hatband[s] & glove[s]	1.0.0
hire of 3 coachmen [&] cloaks	0.3.0
3 hatband[s &] 3 p[ai]r[s of] gloves	0.9.0
feathermen carrying lid, hatband[s] & glove[s]	0.3.0
Man to attend the funeral	0.2.6
Cash paid parish dues of St Dunstan in [the] West	3.0.10
Cash paid mens beer	0.3.0
affidavit	0.1.0
[total]	26.16.7

(Richard Carpenter, Purchase Ledger 1746–7 & Sundry accounts 1761–78, Guildhall MS 5871)

This bill resembles the account of the somewhat smaller funeral of Sarah Hurlin (Appendix B1) which took place at Christ Church in 1839. A far greater diversity existed between the social classes, according to what the undertaker, for the most part, dictated as 'proper'. Mr Wild, an undertaker interviewed by Chadwick (1843, 50), thought that the funeral of a 'tradesman of the better class' should cost between £70 and £100. He felt that £150 for a gentleman's funeral was a low average, and that between £200 and £1000 was what he would expect. A child's funeral of the latter class, however, would only cost about £50.

4 Death and disposal in Spitalfields

The death of an individual who was to be buried at Christ Church was the beginning of a series of environmental and human-derived processes which have to be identified and considered in order to understand the archaeological data. These processes will culminate, for those excavated at Christ Church in first a palaeopathological study, and then cremation and replacement in the crypt. Chapter 7 forms a discussion of the archaeological perspective on these processes, from the moment of death onwards. Unfortunately archaeological data provide little insight into the crucial period between death and burial for those individuals interred in the crypt, partly because work in this field of study is still in its infancy (Henderson 1987).

In general the socio-historical aspects of the 18th and 19th century funeral are well covered by historians such as Gittings (1984), Morley (1971), and Curl (1972). Part of the justification for this project was the unprecedented opportunity to assess some of the generalizations made by historians in the context of an archaeological and historical assemblage.

For the early part of the burial period at Christ Church the historical documentation consists almost solely of the burial register for the parish. This register does not indicate whether an individual was buried in the churchyard or the crypt. However, it does give the exact date of burial. In those cases where biographical information survives on a coffin plate we can calculate the interval in days between death and burial. This is important not only in assessing the effects of environmental processes on an individual during the period before interment, but also in terms of the social and economic implications of delaying burial. From a sample of 275 individuals for whom this interval can be calculated there is a general trend for the delay to increase during the late 18th and early 19th century, from just above five days to almost eight. The range of delay hidden by that average is considerable, from burial on the day of death up to a period of seventeen days. There were practical reasons, especially during warm weather, for disposing of a body as soon as possible after death, in order to avoid both unpleasantness and danger to health. However, there seem to have been many more reasons for delaying the funeral, as Chadwick noted:

> The causes which influence this practice amongst the greatest number of the population appear to be, first, the expense of the funeral; next, the delay in making arrangements for the funeral; the natural reluctance to part with the remains of the deceased, and occasionally a feeling of apprehension, sometimes expressed on the part of the survivors, against premature interment.
>
> (Chadwick 1843, 46)

This last possibility induced the production of a number of publications, including Joseph Taylor's (1816) *The danger of premature interment proved from many remarkable instances of people who have recovered after being laid out for dead and of others entombed alive for want of being properly examined prior to interment.*

Chadwick saw a class distinction in treatment of the dead. He observed that the 'labouring classes' kept the body for a week, or sometimes two weeks. He thought that there was 'less delay amongst the higher and middle classes' (ibid), and that corpses were placed in lead or double coffins earlier.

At Christ Church the opportunity to compare the archaeological with the historical record does not occur until towards the end of the period of burial in the crypt. There are two strains of evidence, the testimony of apparently objective observers, and that of those directly involved in the burial procedure at the church. Walker noted in 1841 that 'The Spital fields ground ... is literally overcharged with dead. The vault underneath the body of the church is also very much crowded' (Walker 1841, 13). Chadwick, however, was of the opinion that Christ Church was 'one of the best managed palaces of burial in the metropolis' (Chadwick 1843, 100). The two statements read together may simply indicate that on the whole the state of churchyards in the capital was poor. The Reverend William Stone, rector of Christ Church, found his funeral duties altogether too onerous, but offered Chadwick a vivid glimpse of the atmosphere surrounding a funeral at Christ Church:

> Indeed, as my church extends along one side of another crowded street, I have had most inappropriate musical accompaniments, even during that part of the burial service which is performed within the church. My burial ground is partially exposed to the street at the west end also; and there, as at the east, it is liable to be invaded by sounds and sights of the most incongruous description. Boys clamber up the outside of the wall, hang upon the railing and, as if tempted by the effect of contrast, take a wanton delight in the noisy utterance of the most familiar, disrespectful and offensive expressions. ... To this wilful disturbance is added the usual uproar of a crowded thoroughfare — whistling, calling, shouting, street-cries, and the creaking and rattling of every kind of vehicle — the whole forming such a scene of noisy confusion as sometimes to make me inaudible. On all these occasions, indeed, I labour under the indescribable uneasiness of feeling myself out of place. Amidst such a reckless din of secular traffic, I feel as if I were prostituting the spirituality of prayer, and profaning even the symbolic sanctity of my surplice.
>
> (Chadwick 1843, 83)

It is doubtful whether Stone was acquainted with conditions within the crypt itself. The Hurlin letter suggests that at least a part of the burial service might have been read in the crypt. If this was so, then it seems likely that one of the vaults on the south side of the church was used as a vestibule. It is unlikely that relatives of a deceased individual would have been encouraged to view the scene inside the vaults, especially the public, or parochial, vaults. Lewis (1850, 31) suggested that even the sexton of a church was often barely aware of the state of his vaults, the implications of which are considered in section 5 below. It is entirely possible that Stone and his predecessors never themselves entered the major vault areas of the crypt. Indeed, Lewis suggested that sextons had the exclusive access to vaults. This may account for the state of many of them.

Lewis, who himself had not yet visited Christ Church at the time of his report to the 1850 Commission, described an evidently familiar scene in another London crypt. At St Mary at Hill the vaults were in a 'disgraceful condition', of which the church warden was not apparently, aware: 'In one spot thirteen coffins are piled one upon the other, many of them broken and crushed; the bones from the upper coffins dropping down among those of the lower, and mixing with them in all stages of decay, are most revolting to the senses of smell and vision' (1850, 29).

In contrast to the disgust expressed by the incumbent and the 1850 Commissioners, the first-hand account of a funeral which took place at Christ Church in 1839 reflects only the very public aspects of the Victorian funeral. The Hurlin letter (Appendix B1) was written in 1908 by the grandson of Mrs Sarah (Marchant) Hurlin. It describes her funeral, which was held on 17 May 1839, twelve days after her death aged 74. Apart from a very precise account of the procession and the ceremony, Hurlin's description of the coffin is important. This coffin, excavated close to the entrance of the eastern parochial vault, conforms exactly, as far as can be discerned, with the coffin described in the letter. It cannot now be determined where exactly in the crypt was the place 'assigned' to the coffin. After 1844 it was moved from its original place and was one of the last coffins to be placed into the second (eastern) parochial vault before it was sealed in 1845.

Another funeral account exists for Christ Church, recorded in the *Gentleman's Magazine* of January 1806, as part of an obituary for Colonel Paul Le Mesurier. He died on Monday, 9 December 1805, aged 50. The funeral was a lavish affair, as befitted a Member of Parliament and alderman (Appendix B2). It is unfortunate that the vault of the Le Mesurier family has not been identified. The description of the funeral amply illustrates the aristocratic nature of a high-status funeral of this period, in almost comic contrast to the middle-class departure of Sarah Hurlin.

In section 5 below, and in chapter 7, the difficulty of inferring the precise nature of ritual events from the archaeological evidence is discussed. It has profound implications for mortuary archaeology in other periods.

5 Mortuary behaviour: the archaeological evidence

Since (see chapter 7.4) *de facto* refuse was largely absent from the deposits at Christ Church, techniques used by sextons and their assistants for conducting mortuary activities in the crypt cannot be reconstructed using archaeological evidence alone. Analogies for these activities have, however, suggested themselves. The excavators themselves faced the same problems in retrieving coffins as the sexton must have faced in introducing them. There are many techniques for lifting very heavy objects and great bulk which rely on using more work and less force: using wedges and supports to keep coffins off the floor; employing a fulcrum to manoeuvre a coffin round a corner; using supporting slings with four people lifting it. Many methods were tried by the excavators, primarily in order to avoid sustaining injury. As a result, efficient and safe methods were devised which may provide analogies for the ways in which interments were originally introduced. A device observed at Withyam in Kent may have been used at Christ Church. It consisted of a stair-like ladder on which the rungs rotated, like rollers, so that a coffin could be raised from or lowered into a vault by using a single, controlled point of force, like a pulley. Such a device might have been used on the south-west staircase into the Christ Church crypt, in the Peck vault, or at other entrances whose exact form we do not know.

The absence of sufficient coffin furniture deposited as rubbish in the crypt compared to the amount which is 'missing' from coffins is significant. This phenomenon has been referred to as the 'Houdini' bias (Kranz 1974), and its implications for formation processes have been noted below in chapter 7. The 'Houdini' bias was used to describe how fauna might escape from situations of imminent catastrophic burial, thus introducing a bias in a recovered assemblage. In the case of coffin furniture at Christ Church the 'Houdini' effect could be the result of a chemical/mechanical destruction process acting differentially, which seems unlikely, or the result of cultural processes. These latter might include lateral cycling, reuse, secondary deposition outside the excavated area, or even ritual deposition for which we have no evidence. It would seem unlikely that individuals who were involved in a recycling process would be likely to advertise the fact.

It is evident that other funerary material was indeed known to have been recycled during the period 1730–1850: coffin wood, lead, human bodies, funerary clothes. A mechanism for recycling coffin furniture is not known from historical data, but it is by no means impossible that undertakers might provide an incentive for some material to find its way back from the crypt to the workshop.

There may be some direct archaeological evidence for the activities of the 'searchers' hinted at in bills for funerals. These people were apparently hired for relatives by undertakers to ensure that an interment was placed as close to others from the same family as possible.

6 Spatial analysis

In 1981 Goldstein (1981, 51–69) addressed the spatial component of mortuary sites which had previously been neglected:

> ... examination of the spatial component can yield information on at least two broad levels: (1) the degree of structure and spatial separation and ordering of the disposal area itself may reflect organisational principles of the society as a whole; and (2) the spatial relationship to each other of the individuals within a disposal area can represent status differentiation, family groups, descent groups, or special classes, dependent upon the correlation of these spatial relationships with other dimensions of study.
>
> (Goldstein 1981, 57)

It is not within the scope of the present report to analyse the degree to which the Christ Church crypt reflects these components; that must await further work. However, it seems appropriate to outline some of the spatial components of the crypt, and to suggest the bearing that they may have on a fuller understanding of Christ Church as a mortuary site.

Spatial components may be broadly divided into four groups. The first of these comprises the smallest unit of disposal at Christ Church: the body. Its size, weight, shape, and space within a coffin contribute significantly not only to the exact place of disposal of an interment, but also to the processes which may operate on it during and after the time of initial deposition. For example, the size and shape of a coffin determined largely the way in which a vault such as the EP area was used. Adult coffins were placed on an east–west or west–east alignment, allowing two north–south rows of coffin stacks to be created maximizing the use of space. This reinforces the inference that space was the most critical resource controlling the disposal of the dead at Christ Church. Interments placed on an north–south alignment, contrary to orthodox practice, or placed vertically, indicate strongly the overriding concern with maximizing use of space.

The second group of components are those which affected mortuary activity on a larger scale, that of the site. These include the shape and architecture of the crypt, the size of the disposal area, and the constraints imposed on disposal by difficulties of access. The tunnels and chambers in the tower foundation may have shown themselves to be highly convenient for conversion to private family vaults, if they were not actually intended to be used as such from the beginning, despite the Commissioners' ruling. This may have contributed to an apparent distinction between the private and parochial areas. Interior partitions were created in the crypt after the construction of the church, either to create more discrete areas for private burial (which resulted in increased income for the church and its employees) or to ensure convenient access to all areas which were likely to be used. In this context, the larger an area which could be enclosed by a simple brick partition wall, the more likely that it would be employed as a public vault. This may have strong implications for social variability at Christ Church.

The third group of components is represented by the community which the church served both as a place of worship and as a place of disposal. The church itself was designed to be at the centre of the local community, and thus if the church was at the centre, then so was the crypt. This position is reinforced by accounts, such as the Le Mesurier funeral (Appendix B2), which emphasize the church as a focus for communal grief or celebration.

The role of searchers has already been mentioned. It is axiomatic that a desire to be buried as close to one's family or friends as possible, reflected in the activity of searchers, creates spatial elements which are socially and archaeologically significant. It is important to note that this activity must have resulted not only in clustering of familial groups, but also in the displacement of other patterns of interment during rearrangement. There is no evidence, however, that breastplates or endplates may have been moved from coffins instead to give the illusion of family clusters.

The church as a corporate body could also influence spatial patterning within the burial area. The tradition of areas of desirability, near the altar for example, could explain the choice for the location of some of the private vaults. The employment by the church of a dissolute sexton and assistants could also affect the final positioning of a coffin within the burial space. However, archaeological evidence that any of these processes may have taken place is indiscernible, although it would be very significant if there were any means of inferring it.

The fourth element in the spatial matrix is the effect produced on the mortuary environment of non-cultural or environmental transformations, such as graviturbation (see below, chapter 7.5; Adams & Reeve 1987, 252) and faunal activity. Assessing the extent to which these transformations can alter and have altered the specifically spatial component of the archaeology remains a long-term problem.

7 Social variabilty

Social variability tends to be the formidable fence at which many otherwise laudable mortuary studies fall. Christ Church is likely to be no different, except in one respect. The high level of inference obtained at this site means that some attempt can be made at interpreting this difficult area.

That aspects of life are reflected in death there is no doubt. Indeed, part of the justification for investigating so recent a burial population is in the possibility of resolving some of the issues which hinder the development of this area of study. For the present its main contribution must be limited to the role of a cautionary tale. It is, however, possible to indicate some of the directions along which analysis may proceed. Christ Church may, for example, play a role in the generation of hypotheses relating to the use of disposal locations in controlling corporate resources (Saxe 1970; Meggitt 1965).

O'Shea (1984, 1) notes that there is 'a common intuitive feeling among all archaeologists that they are somehow very close to the essence of a past culture when working with burial remanins'. But he points out that despite an acceptance of the theory that an individual's treatment in life is at least broadly reflected in death, 'a comprehensive archaeological theory of mortuary variability has not yet emerged' (*ibid*, 3). It is clear that the Christ Church study will play a role in the development of such a theory. Not only are the components of social differentiation and complexity reflected by the crypt burials relatively well understood in the context of the living society which they represent, but the processes of formation of the deposits offer a very high level of inference (see chapter 7). Christ Church may also offer a large number of data pertinent to the study of archaeological impacts. All archaeological data are produced by impacts, although studies of impacts have hitherto largely concentrated on their negative aspects (eg Schiffer 1987; Wildesen 1982). The quality of the archaeological context at this site offers the chance to concentrate on the positive aspects of impact theory.

5 Material culture

1 Background

> The general impression, including the 'official' view of central government and local museum sponsorship, was that we knew all we needed to know about Christian burial, that in any case they had no finds...
>
> (Watts & Rahtz 1985, 128)

Until the excavation of the crypt at Christ Church, Spitalfields, funerary artefacts of the 18th and 19th century were not regarded as material worthy of intensive archaeological study. Archaeological investigation of Christian funerary practice and art has until recently been considered peripheral and unnecessary in a culture which is still ostensibly Christian. There also exists a genuine feeling that to examine the graves of individuals who can be demonstrated to have living relatives is unseemly, even indecent. These types of concerns rarely surface for archaeologists when they excavate the remains of their Anglo-Saxon or Roman ancestors. As a result, discussion of Christian burial culture of the post- medieval period is scant in the archaeological literature, limited to a few site reports where often the only evidence consists of a collection of the 'finest pieces' of coffin furniture.

A notable exception is Rahtz's discussion of the artefacts of Christian death (1981). The discussion includes the wide variety of grave structures and body containers which can be found in Christian contexts and their changing iconography and indicates that study of the Christian dead has potential for anthropological research (Rahtz 1981, 135). Litten (1985), in a much-needed assessment of the nature and contents of post-medieval burial vaults, describes the components of an 18th or 19th century coffined burial. Boore (1985) gives details of the 17th–19th century burial vaults excavated beneath St Augustine the Less, Bristol, and refers to the coffins and coffin furniture of the period. The experience, quoted above, of Watts and Rahtz (1985), demonstrates the problem facing archaeologists with post-medieval interests, forced as they often are to collect inadequate samples of material culture during the hurried removal of the burials to which they belonged. There are other references to the artefacts of Christian death of this period in the archaeological literature (Bishop 1978; Butler 1978; Caldwell 1976; Redknapp 1984; Rodwell 1981; Shoesmith 1980; Thompson and Ross 1973), and examples of some of the more spectacularly fashioned coffin furniture have been preserved for display in museums and art galleries.

Although there has been comparatively little interest in the style, symbolism, or manufacture of coffin furniture of this period in the modern archaeological world it was the subject of heated debate in the 18th and 19th centuries. The industry came in for both economic and aesthetic censure. As early as 1721 the Reverend Thomas Lewis complained that 'gain and lucre' were the motivating factors behind the rapidly growing commercial undertaking trade (see above, chapter 4). However, it is clear that undertakers themselves made little, if any, profit from coffin furnishings: indeed, they often appear to have made a loss on that part of the business. By the middle of the 19th century the term 'coffin-furniture discount' was apparently proverbial (Aitkin 1844).

Suppliers of furnishings largely controlled their style and price; two extant pattern books illustrate the range and cost of items which would have been available to the funeral trade (*Pattern books* 1783 and 1826). Items were sold by the dozen or in sets, which would typically comprise a breastplate, two lid motifs, and three or four pairs of handles. Purchasing by the dozen was paradoxically more expensive than buying by the set. By this means the wholesaler could control lines of stock and ensure that even unpopular and dated designs were profitable, by including them in sets with otherwise popular designs. For the archaeologist or art historian who relies on a concept of fashion in taxonomic theory this economic reality is rather disturbing.

These and other aspects of the funerary industry worried the self-appointed guardians of the British moral heritage. The Oxford Movement of the early Victorian period had begun to force a reassessment of the imagery found in church buildings, graveyards, and on funerary trappings. The Ecclesiological Society, formerly the Camden Society of Cambridge, provided an arena for the so-called 'battle of the styles'. Much of the symbolism which was criticized, particularly Carolean and Rococo motifs, had itself been inspired by a reaction against a prevailing school of expression, 'the ordered world of orthodox Palladianism' (Crook 1978). This was often combined with older, mainly classical imagery. However, a major criticism of this style was that it was un-Christian and that images such as puffy-cheeked cherubs' heads, inverted torches, and weeping women clutching at urns were pagan.

A W Pugin, a staunch advocate of the return to Christian symbolism, wrote *An apology for the revival of Christian architecture in England* in 1843, in which he recommended that 'the essentially pagan . . . urns, broken pillars, extinguished lamps, inverted torches and sarcophagi should have been substituted [by] recumbent effigies, angels and emblems of mercy and redemption' (quoted by Morley 1970, 54). Aitkin, in his assessment of the coffin furniture industry in Birmingham, railed against 'this miserable desecration' (1844, 706), and laid the blame for the inclusion of symbols such as 'this tall damsel, trumpet in hand, about to announce the crack of doom thereon, [and] — this disconsolate, but no less classical matron, embracing the urn over which the cypress, if indeed it be not a weeping willow, drooping so impossibly' (*ibid*, 706) squarely on the shoulders of the architectural world: 'Who shall tell how architecture — classical, gothic and barbaric — has been ransacked to furnish the ornamentation of coffin breast, foot and handle-plates ?' In an article published in *The Ecclesiologist* in September 1844 the extent of moral self-righteousness displayed by the standard-bearers of Christian symbolism was shown by a reluctance even to record the name of the deceased, if they had come from a humble background, but that the exclamation 'Jesu mercy' would suffice. It was further stated that the mention of occupation on a commemoration was entirely inappropriate except in the cases of ecclesiastics (Morley 1971).

It is evident that the study of a large corpus of material from this period not only offers the chance to create a firm taxonomic base for the further study of later Christian funerary artefacts, but may also play a part in creating a context for the historical study of religious and moral reform in the 18th and 19th centuries.

2 The burial container

A comprehensive synthesis of the English funeral, including a historical survey of all aspects of coffins and coffin furniture from 1530–1850, has been compiled by J W S Litten (Litten 1991). The scope of this chapter will therefore be limited to a survey and discussion of the coffins and coffin furniture excavated from Christ Church. All coffins will be referred to by their burial number, or by the first burial number if more than one exists, for ease of reference to the Burial catalogue (MF: M4).

The majority of containers were flat-lidded single-break coffins (using the terminology of Litten 1985, 11), the familiar 'coffin-shape' design which is still used for many coffins in Britain today (Fig 5.1). These were constructed from a combination of elm, oak, conifer, lead, and iron. The burial container frequently consisted of a single coffin shell made of wood. Sometimes an internal wooden lid was added. A 'Russian Doll-like' coffin comprising up to three coffin shells made out of wood and lead was also common. Four burials did not conform to the single-break design and 22 had no signs of a burial container when excavated. The four burials which did not conform were all made from wood, and three of them also had inner wooden shells: one was trapezoidal in shape (2213, type 62, see below), being wider at the head end and narrower at the foot end, and three were rectangular in shape (5003 and 2451 of type 42, and 2060 of type 40, see below). Two of the rectangular coffins were notable because they contained multiple burials; one (5003) contained parts of at least 19 individuals and the other (2060) 3 adult bodies: all three coffins had been braced with iron straps. Table 5.1 shows the variety of facilities in order of frequency, their type numbers, and the percentage of those dated with their known date ranges.

The date ranges for each category show no significant trends for types 01, 03, 38, 66, 31, 36, and 37, which represent 94.9% of the total sample and are spread evenly throughout the period of use of the crypt. Categories 35 and 79, which represent only five burials, show two shorter date ranges. Type 35 indicates a burial which consisted of a single inner lead coffin representing the remnants of a type 36 container, which had lost its surrounding wooden shell in a redeposition process. The close date range of this type is therefore not considered significant. However, the two burials (2437 and 2438) of type 79 are interesting because they consist of outer lead shells which encase decorated wooden shells containing plain inner wooden coffins. The decorated wooden shells appear to have been intended for use as outer wooden shells: they both bore breastplates and one, 2437, had an endplate attached to it (an endplate [ENPL] is a small lead plaque thought to have been used to identify coffins whose breastplates would have been obscured when they were stacked in piles in the crypt). Both of these burials postdate an order of the Vestry in 1813 that for hygienic reasons all further burials should be encased in lead. In the eastern parochial vault, where they were both located, 89.3% of all interments were encased in lead. It is therefore quite likely that the outer lead shells were added to comply with the Vestry rules.

All the single-break coffins had been constructed using butt joints, which in adults' coffins were fixed with small nails or panel pins, sometimes after the internal surfaces of the joints had been glued together; the lid was then bolted into place with bolts at the corners and at the shoulders. The use of such bolts was not observed in infants' coffins. Most were built up from a single-break-shaped base onto which the two single-pieced sides were fixed. The sides were kerfed (that is, scored vertically 6–7 times internally at the shoulders), steamed into shape and then fixed around the head and foot boards. Some of the inner coffins deviated from this design by having the sides of the coffins fixed onto the base between the head and foot boards. Generally the head board was fixed vertically onto the base, although the foot board was often set at an angle of more than 90 degrees to the base (see Fig 5.1). If an extra internal lid was added to the design, the edges of it were normally chamfered and set

COFFIN CONSTRUCTION

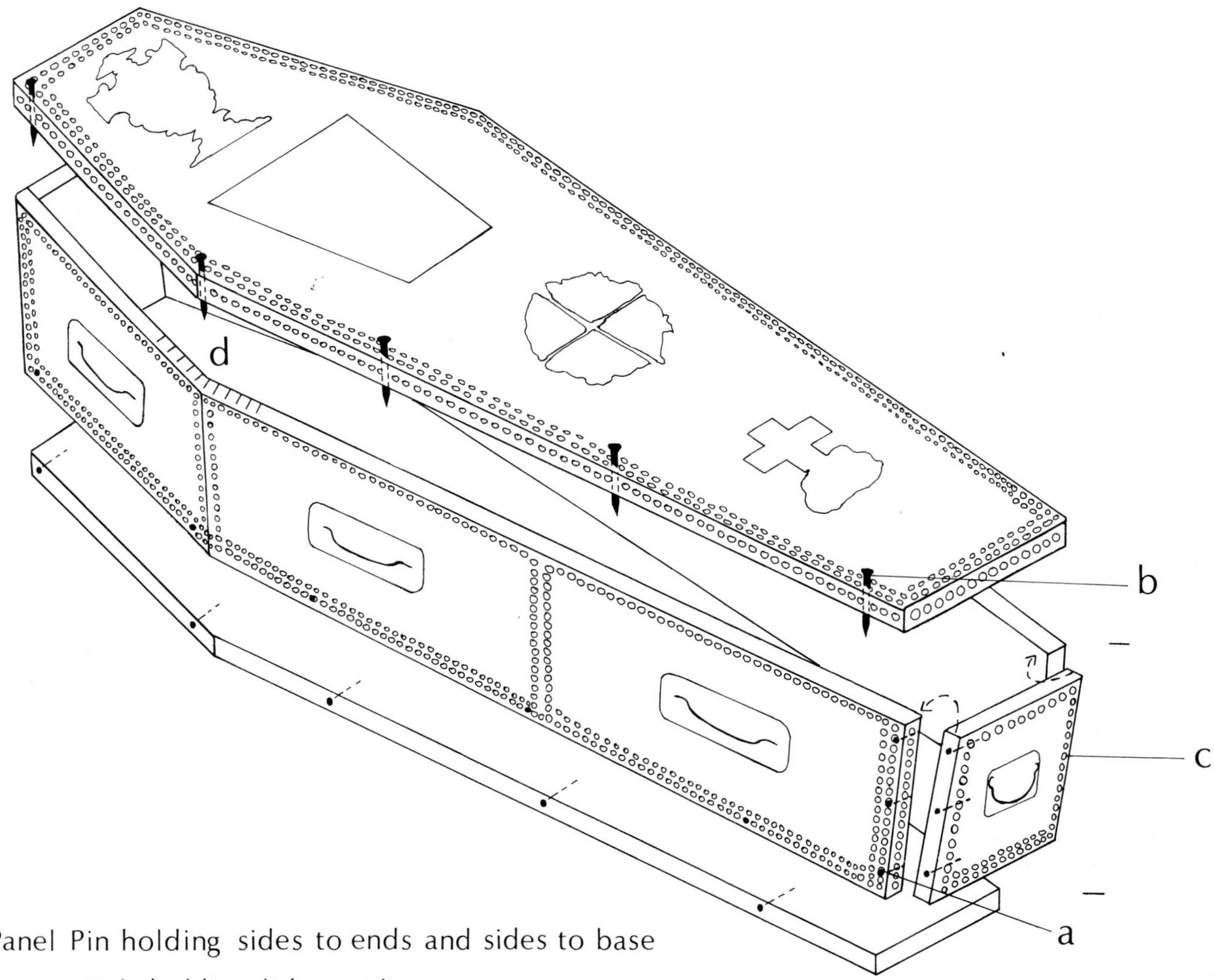

a: Panel Pin holding sides to ends and sides to base

b: Corner Bolt holding lid to sides

c: Upholstery Pin holding covering to sides and lid – usually decorative

d: Kerfs – cuts to allow bending of side

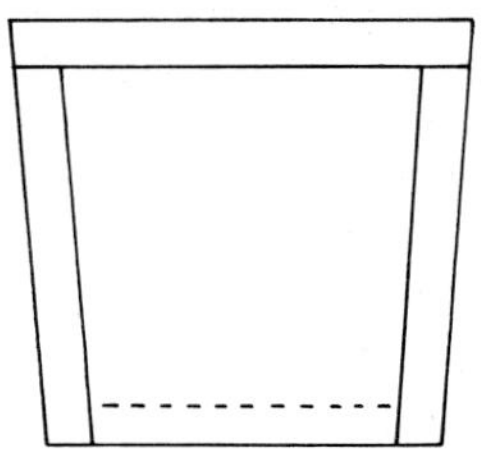

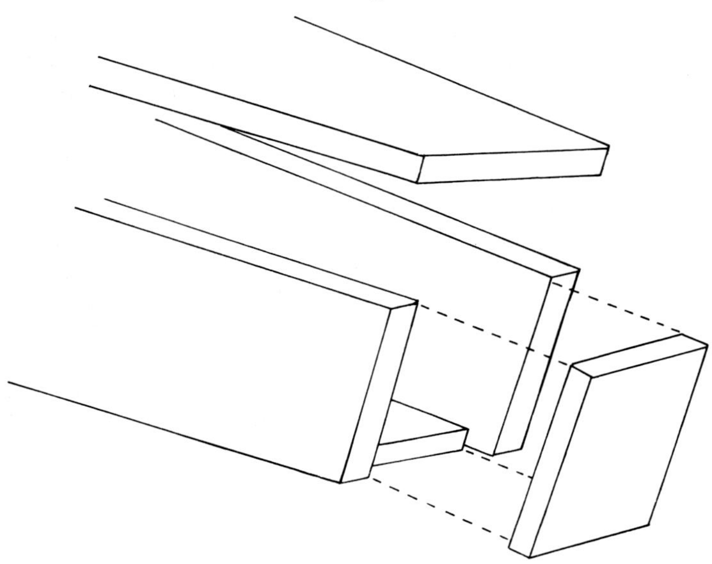

Butt Joint: sides outside ends and base but inside lid.

Figure 5.1 Exploded view of a single-break coffin

Table 5.1 Details of variety of burial containers at Christ Church

Type	Shells	Outer	Inner		%of dated ones	Date range
01	1	W/I			087/442/19.7	1741–1835
03	2	W	W		039/152/25.8	1746–1847
38	3	W	L	W	117/125/93.6	1732–1852
66	3	W	L	WL	106/124/85.5	1743–1845
31	2	W	WL		023/065/35.4	1729–1839
55	0				000/022/00.0	–
36	2	W	L		016/021/76.2	1734–1834
02	1		W		000/012/00.0	–
65	2		L	WL	000/005/00.0	–
37	2		L	W	002/004/50.0	1793–1826
35	1		L		002/003/66.6	1806–1843
79	3	L	W	W	002/002/100	1815–1825
62	2	W	W		000/001/00.0	–
42	2	W	W		000/002/00.0	–
40	1	W			0000/001/00.0	–
45	1	WL			000/002/00.0	–

W = wood; L = lead; I = iron; WL = wooden lid

with the chamfer uppermost onto a thin strip of timber beading which had been tacked on to the inside of the coffin *c* 40 mm below the top edge.

There was no apparent significance in the differentiation between the species of wood used to construct outer and inner wooden coffins (see Table 5.2). The most frequent thickness of wood used was 25 mm, but some of the wood used in the construction of outer wooden cases had been cut to a thickness of 70 mm. Wood for the construction of inner cases was sometimes thinner than this at 20 mm. All the outer wooden cases of burials were found to show vestiges, at least, of having been covered in upholstery and none of having been finished by polishing or waxing: the wood would not have been seen at the funeral. Some of the lids and bases from both outer and inner wooden coffins had not been cut from a single piece of wood but created by fixing offcuts together with panel pins before the shape was cut out.

The trapezoidal coffin (2213) was remarkable for the different joining techniques used in construction. The lids of the outer and inner wooden cases had each been made from seven pieces of wood fixed laterally, with tongue and groove joints in the case of the outer lid, and with butt joints and panel pins in the case of the inner lid. The corners were fixed with a transverse nail across the corners, which had been mitred. The rectangular coffins, 2060 and 5003, were constructed out of planks of wood using butt joints and panel pins: two planks per side, two per base, and two for the lid, with corners braced internally by wooden blocks. The double-shelled coffin (2451) was a similar shape to 2060 from the foot to the shoulder. However, the sides had been cut from one piece of wood, and additionally at the shoulder of the coffin an extra piece of wood had been nailed on, creating a slight angle in the side and making the head end of the coffin narrower than the foot end. All three of these rectangular coffins were reinforced with iron bands which had been nailed on to

Table 5.2 Wood types used in construction of outer and inner coffins

Wood type	Outer coffin	Inner coffin
Elm	411/553/74.3	158/185/85.4
Oak	027/553/04.9	002/185/01.1
Conifer	115/553/20.8	025/185/13.5

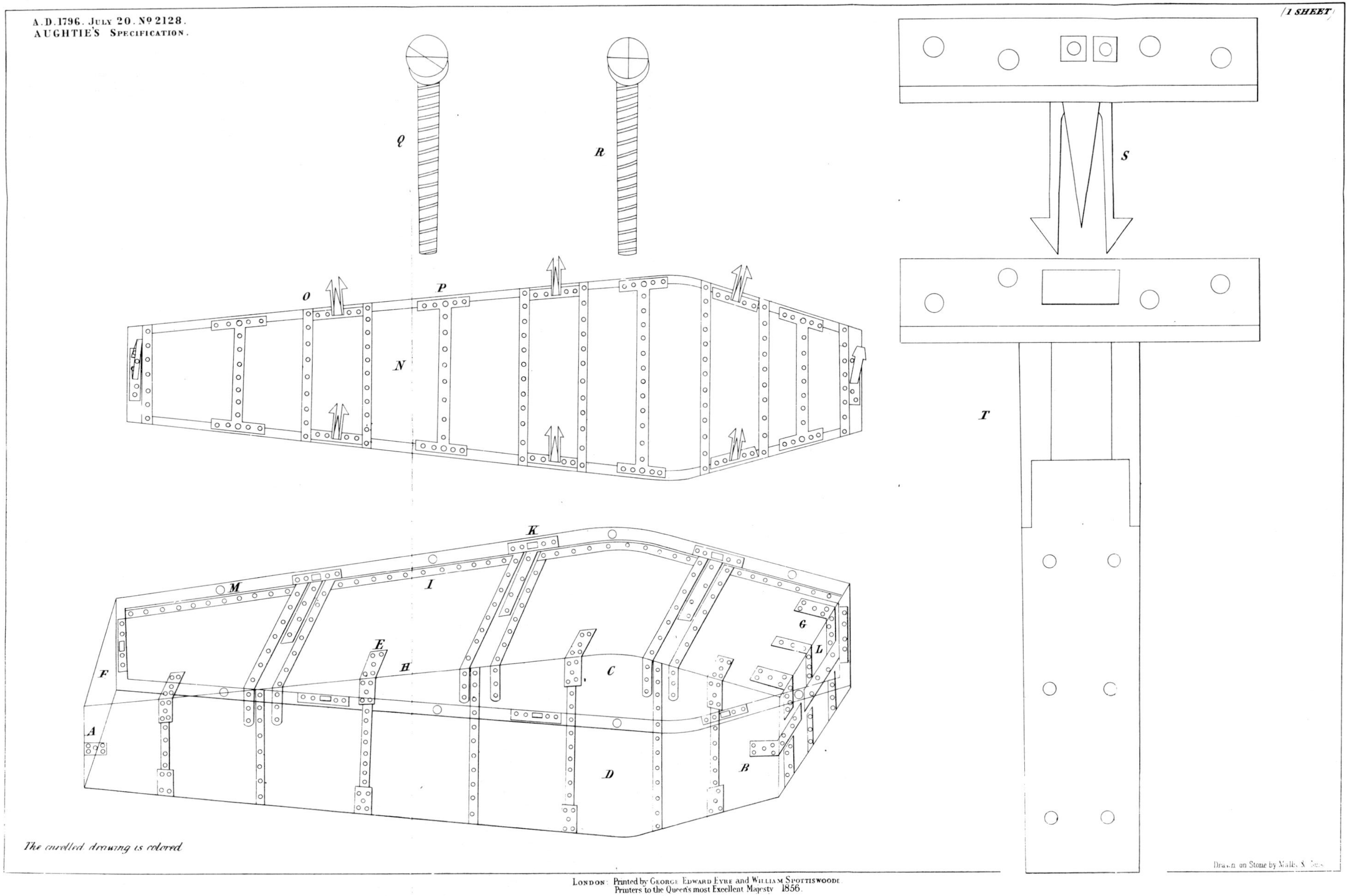

Figure 5.2 Patented coffin of Gabriel Aughtie

Table 5.3 Details of the lead construction designs at Christ Church

Type	% of dated ones	Date range
01	010/011/90.9	1739–1847
02	088/102/86.3	1732–1849
03	006/006/100	1775–1830
04	001/002/50.0	1825
05	010/014/71.4	1770–1844
06	002/002/100	1779–1831
07	046/052/88.5	1749–1852
08	052/057/91.2	1737–1852
09	017/018/94.4	1758–1843

the outside of the coffins after they had been closed. There were 27 other incidences of single-break coffins being reinforced with iron bands and one case of a chain being fixed around the outer cases of a burial (2484). Three of these were dated — 2301, 2418, and 2484 — and also had a lead shell, making the bodies contained within very secure indeed. None of the iron-banded coffins rose to the intricacy of the patented coffin of Gabriel Aughtie (Fig 5.2), but they might have provided 'A coffin so secure as to render it impracticable either to break, cut or otherwise open the same, and consequently to prevent the stealing, removing, or taking away the body or bodies contained therin' (Aughtie 1796). Fear of 'resurrectionists' at the end of the 18th and the beginning of the 19th century are discussed in chapter 4 (p66).

In 19 instances carpenters' marks were observed. The possible meaning for 18 of these has not been discerned, but the marks on 2123 written in white chalk have been identified as a measurement. The significance of the numbers '5.6 19' was pointed out by a contemporary undertaker: they note the internal length (in feet and inches) and shoulder width (in inches) of the coffin (I Leverton, pers comm). These were chalked on the outside of the outer coffin side, possibly while it was waiting in the undertaker's workshop or yard to be used for an appropriately sized body.

The insides of both inner coffins and single-shell wooden coffins were sometimes pitched. The pitch might cover the entire base of the coffin and all the internal joints, or just the foot end of the coffin. This was probably a precaution against leakage.

Nine different variations of lead coffin construction were observed (Fig 5.3 and Table 5.3). These can be divided into three major forms, which are referred to here as 'pie-crust' (types 01–04), 'shoe-box' (types 05 and 06), and 'flush-soldered' (types 07–09), terms which describe lid-form. Sable Plume remarks on two major methods of making a lead coffin: a lead coffin could be made to line the outer wooden coffin, a procedure he suggested had gone out of fashion by the end of the 19th century, or a lead coffin could be constructed to surround the previously constructed, and presumably occupied, inner wooden coffin (Sable Plume 1910, 57). In some of the burials at Christ Church which contained lead coffins of the pie-crust design fragments of waste solder were found trapped in between the kerfs of the outer wooden coffin. It may be inferred that the lead coffin had been constructed within the outer wooden coffin (Bowis 1985, 14). In other burials of the pie-crust and shoe-box designs nails had been punched through the lead coffins from the inside to facilitate the fixing on of the outer wooden coffin, implying that the second type of procedure mentioned by Sable Plume was being used. Lead coffins of the flush design seem most likely to have fallen into the second category as well. If an inner wooden lid was included in the facility it was held in place by the creation of ridges on the inside of the coffin, in reality small dents made by a hammer from the outside of the coffin. An assessment of the date ranges for each type of design does not give any significant indicators as to the relative popularity of those designs through time. Many of the lead coffins had been 'decorated' by cross-hatching the external surfaces; Sable Plume suggests that it was done with a 'shave-hook' (*ibid*, 62). As the lead coffin was destined to be an internal feature of a burial (except in facility type 79), it seems to have had no particular purpose.

Two other uses of lead in the burial facility were observed in burials 2561 and 2168. In 2561 a single sheet of lead was used to line the outer wooden coffin; the piece of lead was bent to accommodate the two bottom corners of the coffin, but no ends or lid were added to make it into a complete lead shell. In the case of 2168 (Fig 5.4) two sheets of lead, 0.45 x 0.20 m, were laid longitudinally over the top half of the body. The lead which was used was considerably thinner than the lead used for manufacturing lead coffins. No explanation has been offered for this procedure.

One coffin was made from iron (2247), which contained part of a locking mechanism; the lid was missing. This has been the subject of a detailed metallurgical study which is summarized in Appendix F.

Sawdust was probably used to prevent the body from moving around the coffin and was observed in 235 cases (see chapter 7.4 and 5, for further discussion of sawdust). The quantity of sawdust used varied from a thin layer sprinkled on the base of the coffin to cases where the coffin was filled to the brim. The use of sawdust was not limited to certain types of facility. In one instance, 2696, where sawdust filled the coffin to the brim, a section was divided off by the inclusion of a small open-topped wooden box just above the feet. The box was rectangular (0.24 x 0.18 x 0.095 m), made out of conifer 5 mm thick at the base and 15 mm thick in the sides. It was also filled with sawdust, but did not appear to contain anything else. There is a possibility that it had been used as a viscera box and that the contents had rotted away, as had the soft tissue on the body itself. However, the sawdust from the box has not yet been analysed to prove or disprove this hypothesis.

3 Coffin furnishings

The external surface of the outermost wooden case received further treatment before it left the undertaker's workshop to become part of the funeral proceedings. All the external surfaces of the coffins at Christ Church were covered in upholstery (see p100) before receiving various metal embellishments. (Metallurgical analysis of these metal artefacts can be found in Appendix E.) These consisted of both functional and decorative pieces of metalwork. The terms here used to describe them have mostly been taken directly from contemporary sources. Upholstery pins are round-headed nails which were used to affix the upholstery covering to the outside of the wooden coffin; usually they were arranged in a decorative way. 'Lace' was the term given to the decorative strip of metal which was sometimes used to perform the function of a series of upholstery pins. Grips, also referred to as handles in the contemporary pattern books, were placed at intervals around the outside of the coffin, normally three to each side and one per end in the case of an adult-sized coffin, but only two per side in the case of a child-sized coffin. These could be used to lift the coffin, but in the cases of adult-sized coffins which included a lead coffin, weighing up to a quarter of a tonne, they are likely to have been purely decorative. The grips on a child-sized coffin would probably also have been decorative. On all the outer coffins at Christ Church remnants, at least, of backing plates were observed behind the grips; these have been referred to as grip plates, although they are also called shields in the contemporary pattern books. These plates offered holes through which the metal clips which held the grips onto the coffin were fitted. The breastplates, onto which brief biographical details were recorded, were embellished with Carolean and classical symbolism. These were fixed on to the lid of the coffin below the shoulder. The outer coffin was sometimes decorated further with small pieces of pressed metal displaying similar types of motifs; these were referred to as escutcheons and were also called 'drops' in the contemporary sources. They were fixed all over the coffin to emphasize the overall design, most frequently placed in corners or along lines already established by the pattern of upholstery pins. Larger items of essentially decorative metalwork displaying this same type of imagery were referred to as lid motifs and were found on the lid either at the head end above the breastplate or at the foot end, or sometimes both. There were occasions where more than two were observed although often none were used. On the lead coffins there would sometimes be a plain inscription bearing biographical details. This could consist either of a sheet of lead, sometimes with a simple border decoration which had been soldered on to the coffin, or simply an inscription engraved on the lid. Other types of coffin plates have been referred to with names which illustrated where on the outer coffin they were found: head plates, side plates, and end plates. These were normally rectangular pieces of lead onto which some or all of the biographic details were recorded. Although there is scope for assessing the designs and symbolism of these pieces of coffin furniture through time to test archaeological methods of seriation, typology and space- time unit concepts, as attempted by Dethlefsen and Deetz (1966, 502), this has not been attempted here. Brief descriptions of the range of symbolism and manufacture are discussed for each type of artefact with statistical summaries, including date ranges for designs for each group of material (Appendix D).

Breastplates, grips, grip plates, escutcheons, and lid motifs could be bought separately from the manufacturers or in sets. On the opening page of the pattern book of Tuesby and Cooper, dated to approximately 1783, written over the breastplate pattern no 51 are the details of what a set entailed:

> Ch'ds Br't Plate Wh't 5/- dz
> D'o with hdles N'o 2 1/8 Sett
> D'o White & Blk Shields 1/10 Sett
>
> N.B. A Sett of Coffin Furniture
>
> contains a Breast Plate Flower Pot
> & Angel, 3 P'r of Handles & Pins to
> fix them.
> If order'd with 4 P'r of Handles &
> 20 Yds of Lace &c the Price is advanced
> in proportion
> Large Wr't Cast Gripes to any of the Setts
> will be 1/- P'r Sett advance
>
> (Tuesby & Cooper 1783, E997)

All the design types mentioned in the above 'Sett' have parallels in the coffin furniture found at Christ Church, and have been found in the same combination. Another 'Sett' from a slightly later pattern book details the combination of a 'Glory and Urn' and four 'Gripes No. 11'; this combination was found frequently in the extant coffin furniture and has been illustrated here with the most complete examples from 2203 and 2295, (Fig 5.5, 5.8). Another set from Christ Church which uses common design motifs in all aspects of the coffin furniture is that represented by the grip, grip plate, escutcheon, and lid motifs of 2205 and the breastplate of 2369 (Fig 5.6 and 5.7).

The source of the coffin furniture at Christ Church is obscure. The actual furniture may have been made locally, but there is no indication where the patterns may have come from except on the two lid motifs from 2260, which both have 'Richards No 1180 April 1842' pressed into the lower part of the design. Items of coffin furniture which have identical designs to those found at Christ Church have been observed by the authors at St Paul's, Shadwell, St John's, Wapping, St Botolph's, Aldgate, and All Saints, Poplar in London, and in the vault at Withyam in Kent; and similar items have been observed

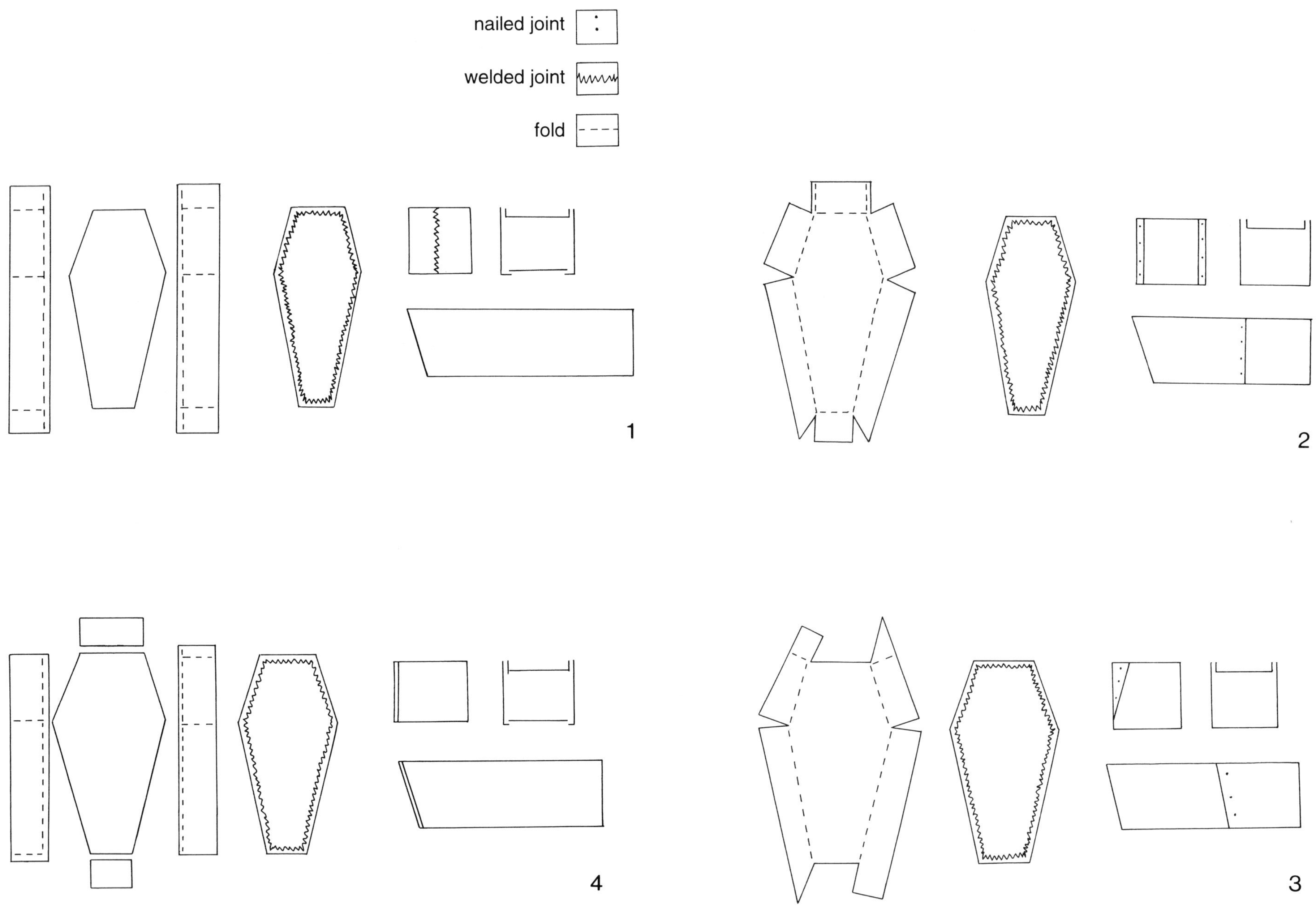

Figure 5.3 Variations of lead coffin construction (1–4)

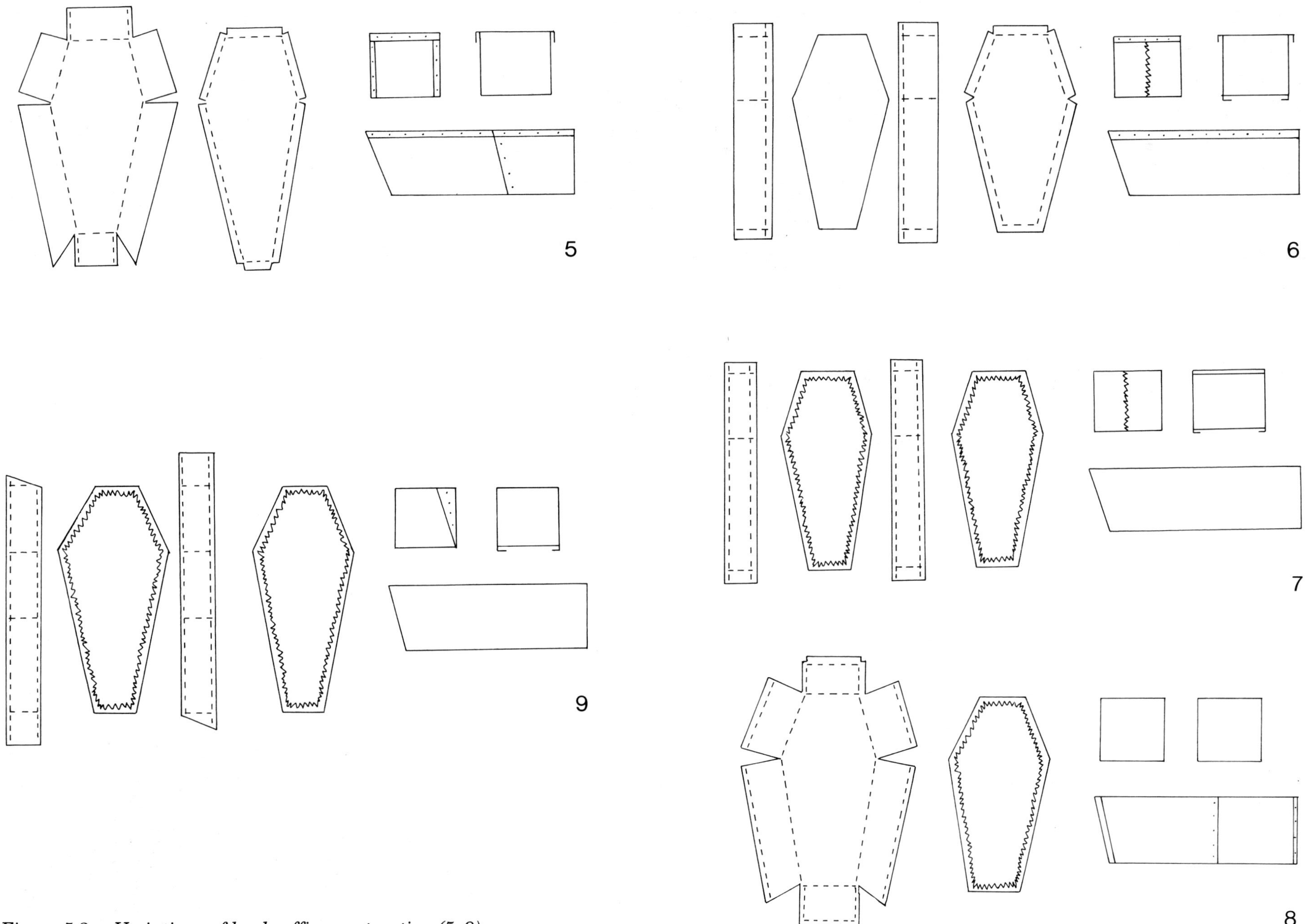

Figure 5.3 Variations of lead coffin construction (5–9)

elsewhere in the country — at St Martin's, Wharram Percy (Harding 1988f), St Mary le Port, Bristol (Watts & Rahtz 1985, 168–75), St Augustine the Less, Bristol (Boore 1985), St Bride's, Fleet Street, London (Gentle & Field 1974, pls 348–66), and at St Mary the Virgin, Little Ilford (Rednapp 1985).

The upholstery pin patterns were obviously used in more than a functional way as 58 variations in the designs they formed on the outer coffin were noted. The pins themselves could be made out of iron or brass and were often covered with black lacquer. It should be noted that an assessment of the apparent fluctuations in popularity of the designs ought to consider both changes in fashion and the differing skills in undertakers' shops throughout the period. No references to patterns for these designs have been found in the literature or in the pattern books. Examples of thin strips of pressed metal skeuomorphically resembling a double row of upholstery pins are to be found in the contemporary pattern books (Victoria and Albert Museum, M63e). These are referred to as nail lace, but only two examples were observed at Christ Church (2307 and 2247). A suggested date range for this type of ornament is given by Gentle & Field (1974: pl 361) as 1780–1810; however, its continued use is mentioned in Sable Plume's instruction manual dated to approximately 1920.

The grips were mostly made from cast iron (see Appendix E) and offered the smallest variation of decorative types, only 12 in total. This may reflect the functional nature of the component. The most frequently observed type, which was used continuously throughout the period, included two winged cherubs as a feature on the lower part of the grip (type 04). It had the highest survival rate in all the differing areas of preservation, due to the durable nature of products made by casting.

The grip plates themselves offered a much greater variation in design and decorative motif: 35 different types (see Appendix D), mostly pressed tin (see Appendix E). The overall size of these components compared to the grips offered a greater area for expression and a much wider range of symbolism and decorative motifs is found (MF M2:B7–C10). Many of the motifs which became the object of argument and condemnation in the 1840s can be observed on the grip plates: winged cherubs, upturned torches, urns, angels with trumpets, and sarcophagi. These often displayed the same motifs which were to be found on other forms of coffin furniture recorded for the same burial. There were some incidences of undecorated grip plates, but these were not common, and again winged cherubs predominated as the major decorative motif.

The breastplates, which were made using various techniques and from a variety of metals (see Appendix E), also offered the greatest range of symbolism and decorational motif: 114 types (MF M2:C11–M3: D9). The most frequently occurring shape of breastplate was rectangular; most of the rest were either trapezoidal or diamond-shaped. However, there were some designs which came into none of these categories: types 35, 36, 38, 39. Some of the decorative motifs were distinctly rococo in style, with degraded versions of flowers, shells and greenery taking the predominant positions on the breastplates (types 03, 07, 08, 14, 15, 22), whilst others, including 61, 57, 112, 10, and 05, included the almost ubiquitous winged cherub motif. Only one design, type 26, displayed the recognizable Christian motif of a cross, although this was in association with the 'pagan' motifs of cherubs, a skull, and a sarcophagus (see Fig 5.7). The biographical information engraved on the breastplates usually followed the form: title, name, date of death, and age at death, eg 'Mr. Thos. Williams died 3rd June 1839 in his 92nd year' (2464). Occasionally 'died' and 'aged' were written in Latin, and often the age was more simply described by the word 'aged' followed by a number.

Figure 5.4 Burial 2168: two lead sheets cover the body (IC.0329.01)

The lettering for the biographical details was mostly simple and undecorated, although there were some incidences of numerals of the year being embellished with decorative designs. The variety of design types overall, and their individual treatment, imply that it was the breastplate, if anything, that the client might specify to the undertaker when arranging a funeral.

Escutcheons, like grip plates, were made from pressed tin (see Appendix E) and were exclusively decorative in purpose; they may reflect a heraldic origin. Sometimes only a small number were used on the lid, placed in the corners of panels, and sometimes escutcheons would cover the entire coffin except the base, in rows reflecting the design of the upholstery pin pattern. These ornaments tended to be small, 4–5 mm across, but still displayed a variety of motif, similar to that of the grip plates. On the whole these tended to be florid throughout the period, although the more geometric sunburst motif only occurred towards the end of the period (see Appendix D).

Lid motifs were predominantly made from pressed metal, although there were examples made from lead and brass (see Appendix E). Lid motifs were observed in fewer cases than escutcheons and because of their size were mostly found singly, although some of the designs did recur in pairs and one design (type 04) was repeated five times on a coffin (2238). The variety of symbolism was similar to that of the grip

Figure 5.5 Lid motif, escutcheon and grips: a typical assemblage

plates, although there was a predominance of angels holding palms, trumpets, the book of eternity, and crowns, often found in association with winged cherubs. One pair of lid motifs which often occurred together combined a flaming urn and two angels holding up a five-pointed crown between them, standing on clouds and three winged cherubs (type 06, see Fig 5.8). This combination was referred to in contemporary pattern books as 'Glory and Urn' because the word 'Gloria' was sometimes written on the crown. As there were relatively few examples of dated lid motifs it seemed unconstructive to assess the frequency of design types through time (see Appendix D).

Plain lead inscriptions were only recorded for burials which had no legible breastplate and therefore the sample observed is more random than that for the other types of coffin furniture. The pieces of lead used to record biographic information on the lead coffins offered very little variation in format. Most were rectangular in shape with no border (type 01, see Appendix D), and the rest varied only in shape — trapezoidal, lozenge- or shield-shaped, sometimes with a stamped border. In nearly half of the observed cases the biographical information was inscribed directly onto the lid of the lead coffin, occasionally also with a stamped border. The format for biographical details was the same as that on breastplates.

The small number of head, side, and end plates attached to coffins suggests that they were not intended as a prerequisite for every coffin. Of these, 72% were observed on coffins in one or other of the parochial vaults; another 22% were noted elswhere on coffins redeposited around 1867. It may be inferred that these types of plates were used as markers in areas which were expected to have piles of coffins, whose stacking arrangements would obscure the information on the breastplates on the lids of coffins. These may have been intended to make the searchers' task of placing relatives next to each other when they were interred easier, if this was requested (see chapter 4). In at least two cases where side plates were present in the parochial vault (PV), relatives had been placed beside each other and on top of each other (2613 and 2192, 2621 and 2737), the latter after an interval of 21 years.

4 Finds from dumps

The only other type of archaeological context apart from burials which has to be considered is the dump. There are three different categories of dump at Christ Church: that representing pre-burial activity; that representing selective covering of burials *c* 1859; and that representing the covering of the burials in the final sealing of the vaults in 1867. The finds are discussed here, not for their intrinsic value or interest, but as indicators for provenances and dates for the actions represented by the dumps in which they were found.

The first type of dump consisted only of bricks, fragments of brick, and mortar. These match the

Figure 5.6 (above and opposite) Collection of furniture from 2369

types of bricks and mortar used in the construction of the church. As all these dumps predate the use of the crypt as a space for burial they cannot give any indications of actions which relate to the burial sequence. They represent building phase activity and therefore should be dated to between 1714 and 1729, the period of construction.

The second type of dump, which was only found in one area of the crypt (SC), consisted of about four and three-quarter tonnes of largely humic matrix with some fragments of broken building material with inclusions of domestic rubbish such as oyster shells and clay pipes. The introduction of this material can be demonstrated to postdate 1852, because it overlies a coffin of that date and predates 1867. After the vaults were officially closed to further burial in 1857, the churchyard was designated to become a 'Lawn or Ornamental Garden' by a church faculty in 1859. This coincided with an order of Queen's Council which stated that coffins should be removed from the vault under the church steps and ventilation holes all around the church should be bricked up. It is possible that the coffins in the south chasm were moved under these auspices and covered with soil from the churchyard.

The third and most substantial category of dump represented about 245 tonnes of material. This covered burials in all accessible parts of the crypt. It consisted mostly of building material, including fragments of church stone- and plasterwork as well as domestic rubbish and rubbish from a leather-working workshop. As the Order in Queen's Council (Appendix A4) clearly states, all unenclosed coffins were to be 'embedded in soil mingled with charcoal'; however, in an urban area in which substantial building work was being done, building material was easier to come by than soil. The church itself had been restored by Ewan Christian in 1866, and fragments of church fabric from the 'restored' galleries and portico areas were found within the dump, suggesting that the 1867 Act offered a very neat solution to a problem of spoil removal. The area was known for its leather working and the inclusion of leather waste suggests that the rest of the 'soil' was also found locally.

5 Artefacts from interments

There were a small number of artefacts found within coffins. Some evidently had been intentionally placed there, some appear to be accidental inclusions, and some were items of jewellery which were adorning the deceased at the time of burial. There were three wedding rings, one mourning ring, one pair of earrings, two glass phials, two wooden trinkets, one tool, and an extraneous glove. The rings were found on the left-hand sides of the bodies near the hands. One wedding ring (2224) was undecorated and made from gold with a maker's mark: 'SG', a date letter 'i', and the head of George III. The date letter signifies 1784 and 'SG' might stand for Samuel Godbehere, who was a goldsmith from 86 Cheapside who worked there from 1784 to 1818. Another wedding ring (2231), also undecorated and made from gold, had the same maker's mark and the date letter 'E', which signifies 1800. A further gold ring (2609) had no marks on the inside and was decorated with lateral grooves around the outside of the ring. The mourning ring (2698) was made out of gold with raised lettering which read 'JUDITH MESMAN OBIT 15TH MAY 1763 AET 29'. The external surface around the letters was covered with blue enamel. The marks on the internal surface were unreadable. This type of jewellery was often given out at funerals to commemorate the deceased (Puckle 1926, 267, pls 268, 270). The two earrings (2251), also made from gold, were two circles, 12 mm in diameter, with two smaller circles attached at the top, presumbably to affix them to the hooks, which did not survive.

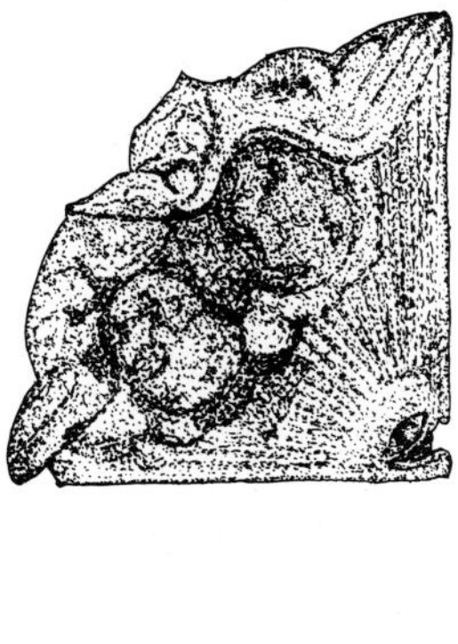

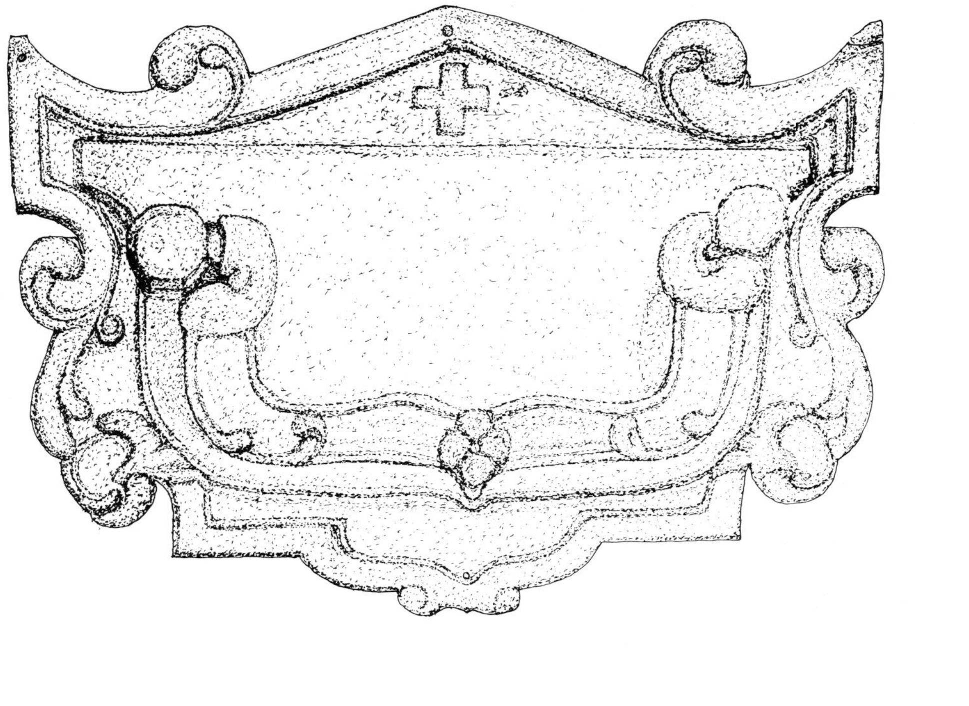

Figure 5.7 Breastplate showing the range of motifs used

Figure 5.8 Lid motif: flaming urn with angels and crown

The two glass phials varied in size, one 70 mm high (2175) and the other 142 mm high (2758), and no longer contained anything, despite the degraded cork still *in situ* in one of them (2758). The small wooden trinket (2715) was a minature turned barrel which could be opened by unscrewing at the middle. Two human teeth were found inside it, one canine and one molar. The molar could have belonged to the occupant of the coffin, but the canine certainly did not. The other wooden object found in a coffin (2705) was a turned annular disk 61 mm in diameter, whose purpose has not been identified. The tool found within coffin 2077 consisted of a toric-shaped wooden handle, 32 mm long, with a protruding shaft of iron extending 87 mm from the centre of the handle. It might be a bradawl, but the iron is too corroded to offer a clear picture. Finally, an extra glove was found in between a clothed body and the wrapping sheet. This might have been an accidental inclusion as it served no purpose in the burial itself.

These artefacts should be considered as a basis for a corpus of dated and closely dated funereal items, not as a definitive taxonomy.

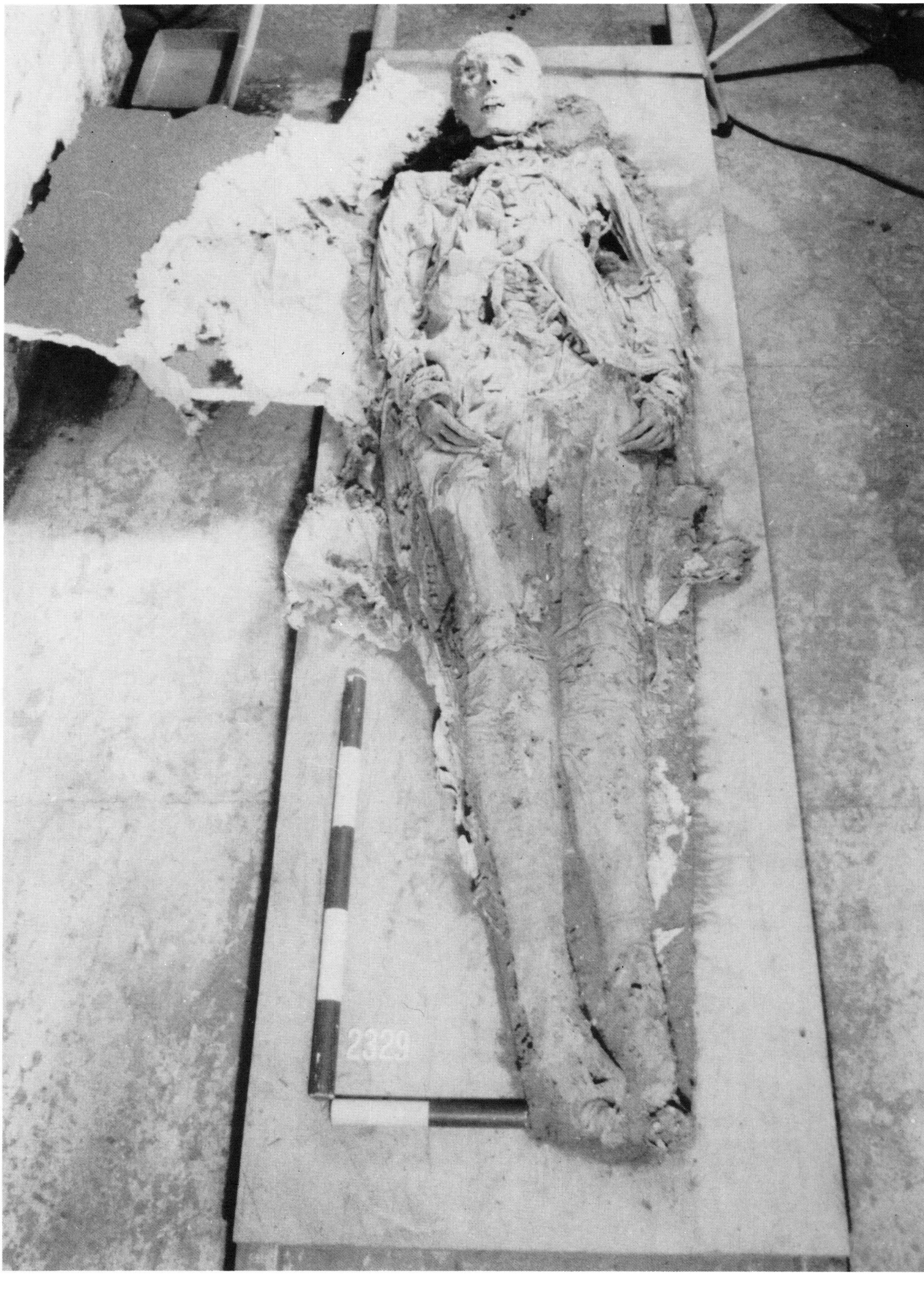
2329

6 The textiles *R C Janaway*

1 Introduction

The excavations produced a large quantity of textile remains, much of it in a very good state of preservation. The textiles can be divided into two groups; those associated with the coffins, either from the inside or attached to the outside, and those found in the dump deposits. The latter group was a mixture of items from broken-up coffins and from the refuse (mostly construction and demolition) which went to make up the dump deposits. With over 500 individual textile finds recovered, it was decided to conserve, but not analyze, the dump material. In addition a quantity of mattress and lining samples will have to await future work. The sampling strategy for textiles changed during the course of the excavation, as a result of both experience gained in recovering and recording material, and changes in priorities (p 26). Initially all textile remains were sampled, which resulted in the recovery of a large number of similar coffin covering and lining samples, evident in the lower burial numbers (see Appendix H, MF). Later, the sheer volume of material led to a policy of taking extensive textile samples only from well-preserved burials, as with the naturally mummified bodies from the north parochial vault as exemplified by burials 2329, 2324, 2325 (Figs 6.1, 6.5, 6.6). Lead coffins in some instances can lead to conditions sufficiently anaerobic to preserve both the textiles and extensive soft tissues, as in burial 2188 (Fig 6.2). In some cases the body was too unpleasant, or considered too much of a health risk, for more than superficial recording, and was sent for immediate cremation. This procedure lead to the loss of some potentially interesting textile items, a loss that can be gauged by the photographs of 2188 or excavators' sketch drawings of 2339 (Fig 6.3). A number of lead coffins with anaerobic conditions were investigated by the British Museum, Natural History Department (BMNH), assisted by Mr A D Mason, who took a personal interest in the textiles. It is these burials (2373, 2407, 2575, 2916, 2917) that produced the best-preserved textiles from the site (Figs 6.8–6.11, Fig 6.14).

In terms of future work on the study of textiles associated with burials of this date it would probably be better to examine a smaller number of well-preserved burials in greater detail than was possible at Christ Church. This would allow not only more careful documentation of garment fastenings but also investigation of problems such as the relationship between soft tissue decay and textile preservation.

2 Health and safety

It was recognized that the major potential source of micro-biological infection was directly from the soft human tissue remains, but the textiles were often heavily contaminated with body decomposition products and so needed careful handling. At the end of the excavation the majority of the textiles were stored with the rest of the artefactual material in a controlled area. Some of the textiles that had been recovered by the BMNH team during post mortems on sealed lead coffins were stored with the human remains. It was agreed that all the textile material should be sent to Bradford University for conservation and analysis, and a policy for transport and treatment was drawn up in conjunction with the project's health and safety advisors. The material, in its original packing, was heat-sealed in a double package of 1000 gauge polythene and marked with the international bio-hazard warning label. Once at Bradford the material was stored and initially examined in a controlled isolation laboratory, although this was considered unsuitable in the long term, because of problems with conservation and access to the material. After taking advice it was considered that after gamma irradiation the material would no longer need to be handled in a controlled area, and more importantly, washing water used in conservation could be discharged down the drains. Test samples were irradiated to assess the level of damage to the already highly degraded fibres, and this was found to be acceptable. All the textile, hair, and leather was sent to a commercial facility for irradiation. At the time this technique had not been used extensively with cultural property. Where it had been used, it was primarily to rid museum objects of insect pests (Urban & Justa, 1986; Justa, 1989).

3 Conservation and analysis

The aim of the conservation programme was to get the material into a condition in which it could be analyzed. Some material, which had been recovered, wet, from sealed lead coffins, was treated as a matter of urgency at the BMNH. The remainder of the

Figure 6.1 (opposite) 2329 body as excavated (naturally mummified) showing dress and stockings

material was stored dry, much of it heavily soiled by decomposition products, the exception being the outer coffin covers, which were affected by dust and corrosion products from the metal fittings. While a few items only required the removal of surface dust with a vacuum cleaner, most items were washed in water and detergent solution. Sometimes several soakings were required in order to remove all the dark brown decomposition products, which released a strong odour on rehydration.

Analysis consisted of preparation of microscope slides for identification of fibre type, and recording fabric structure, construction, decoration, and fastenings. The results are presented in MF M1: A4–B1.

4 Funerary textiles in 18th and 19th centuries

The textiles deposited in the vault, in association with the body/coffin, had three basic practical functions: (1) to clothe the body (2) to finish the interior of the coffin (which would have been open for viewing the body) (3) to finish the exterior of the coffin.

Role of undertakers and funeral furnishers

During the 18th and 19th centuries there were two professional groups who could provide funeral materials and services for non-aristocratic burials. The undertaker was a coffin-maker and arranger of funerals while the funeral furnisher bought in ready made coffins, dressed them, and also arranged the funeral (Litten 1991, 26). 'There was little individual choice of coffin, linings or handles; all was left to the aesthetic judgement of the undertaker, and the client was rarely asked to express his views' (Litten 1991, 30). The laying out and washing of the body (termed dressing and trimming) was a service performed since the early 17th century by funeral professionals who retained the services of number of women to perform this task.

The body was washed and usually dressed in textiles provided by the undertaker. It was relatively unusual for the deceased to be buried in their own

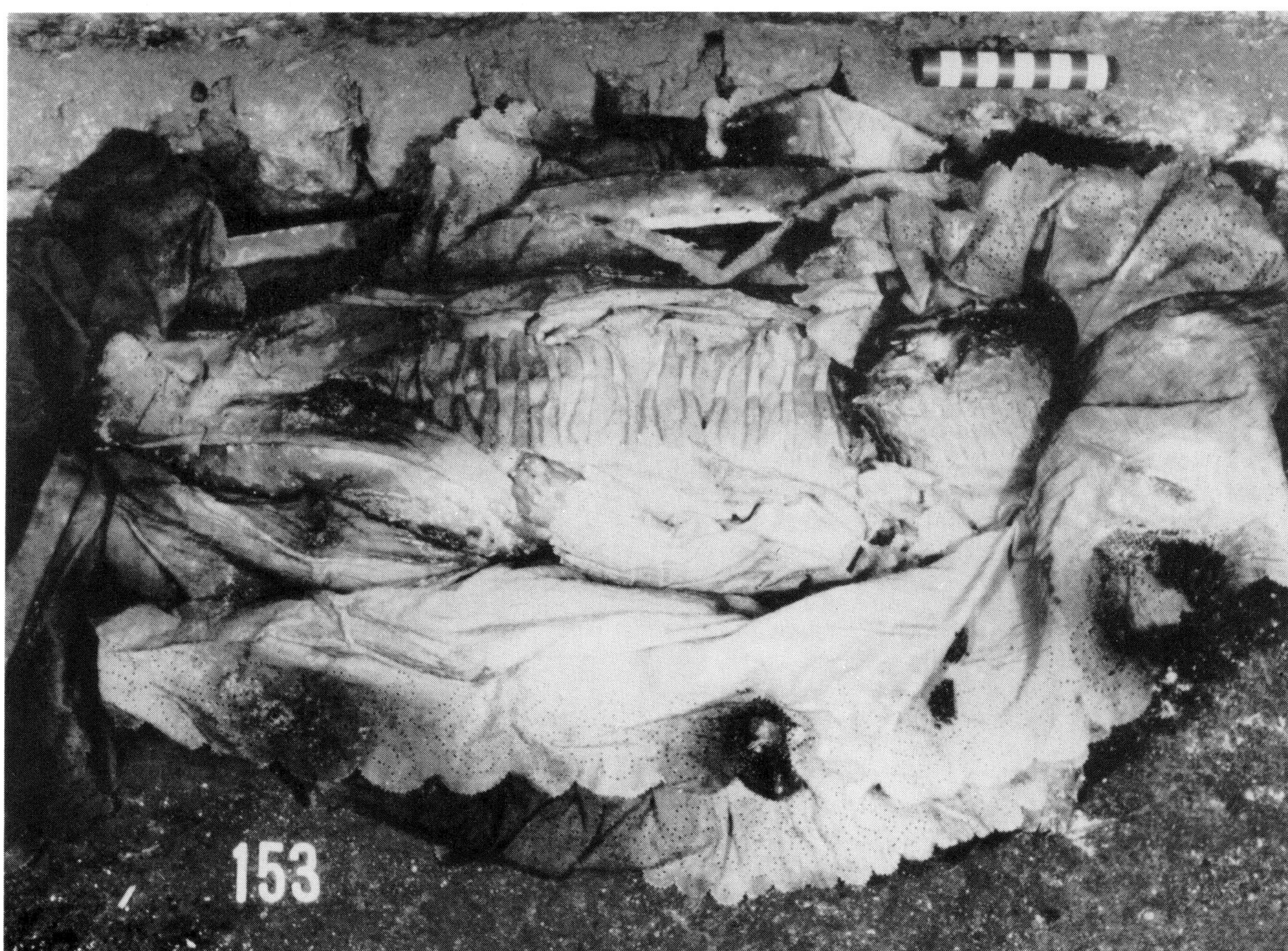

Figure 6.2 2118 baby's body as excavated (from sealed lead coffin); the shroud sheet has been folded back to reveal shroud and cap

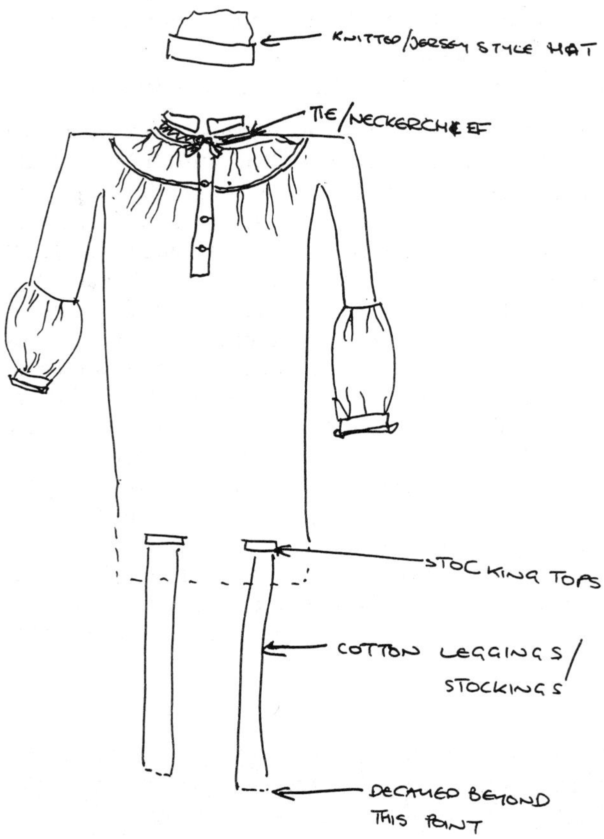

Figure 6.3 *2339 excavator's sketch*

clothes. At Spitalfields 16 bodies dressed in their own garments, sometimes under a shroud, were recovered, while 12 bodies with only shrouds were recovered.

Since the abandonment in the late 17th/early 18th century of the late medieval type of winding sheet (an enveloping shroud which effectively trussed the body), the limbs were no longer secured. This resulted in the practice, which started in the 18th century and continues today, of holding the limbs in position with tapes or ties. The hands could be pinioned against the sides of the corpse with the waist band of the shroud, with the ankles or big toes tied together.

During the 18th and 19th centuries it was normal practice for the coffin to be kept open at home for two or three days allowing friends and relatives to view the body (see chapter 4). This clearly necessitated careful presentation of both the body and the coffin interior as achieving a successful effect was desirable for both the mourners and the undertaker's professional credibility.

Act for Burial in Wool

During the 17th century the choice of material used for shrouds acted as an indicator of status, with different linen cloths being favoured. In 1660 an Act of Parliament was passed in order to promote the domestic woollen industry; this decreed that all persons had to be buried in shifts, shrouds and winding sheets made of woollen material, rather than linen, and free from 'Flax, Hemp, Silk, Hair, Gold or Silver or other than what is made of Sheep's Wool'. This Act, strengthened in 1678, remained on the statute books until 1815, although of course the very wealthy could, if they wished, pay the £5.00 for default. There is some evidence that this practice did occur (Litten, pers comm). One effect of the Act was to encourage the practice of grave clothes and coffin linings being supplied as sets by specialist workshops. This trend is most noticeable with the Spitalfields material, where it was usual for any one burial to include shrouds, caps and other grave clothes made of the same fabric with the same punched decoration as the textiles furnishing the coffin interior.

Coffin construction and outer cloth covering

During the 18th and early 19th centuries outer coffins were usually of elm wood and covered with wool baize or silk velvet. The use of cloth coverings for coffins declined with the introduction of french polishing in the second quarter of the 19th century. Silk velvets were high status coverings and are usually associated with aristocratic burials. They were either black or scarlet in colour during the 18th century with midnight blue, holly green and turquoise being added in the early 19th century. Woollen baize appears to have been either black or scarlet (Litten 1991, 112).

Coffin linings and other textiles used in furnishing coffin interiors

The primary linings were on view to the mourners when the coffin was open. The wood on the inner coffin was caulked on the base and sides with Swedish pitch and an undersheet tacked in place before fixing the side lining, the side sheets and ruched frill. The side linings covered the bare wood, while the ruched frill finished off the top of the lining, masking the beading on which the recessed lids of the inner wooden coffins rested. This helped to form a seal when the lid was screwed down (Fig 6.6). On the base of the coffin a layer of bran and sawdust was placed to absorb liquids and thus help keep the body sweet. A tailored mattress was sometimes used instead of an undersheet and bran. Pillows could be used to support the head.

Winding sheets in the 18th to 19th centuries consisted of large flat sheets, often with decorative borders in punched work. These were laid in the base of the coffin with the sheets extending over the sides and the body, clothed in either garments, shroud, or both, was placed in the coffin, and the winding sheet folded over to cover the body. Litten uses the term 'coffin sheet' for these items. Some confusion however is caused by trade directories of the 19th century still referring to 'winding sheets'. For example, the trade catalogue of J Turner, Funeral Furnisher of Farringdon Street, London (1838) quotes for the following textile items: Shrouds, Ruffling, Winding Sheets and Tacked Matrasses (sic). Other textile materials from his catalogue include Palls, Scarf and Hood, Cloak and Band, Silk Band, Gloves, Velvets for Hearse, Horse Velvets, Hammmercloth.

Side sheets allowed undertakers to dispense with full length shrouds.

> Once the coffin had received its primary lining and edged frill, two rectangular sheets, both the length and width of the coffin were tacked to the base at its sides. Once the body had been placed in the coffin and the fitted pillow positioned under the head, these sheets were then folded over the remains and either pinned down, or roughly sewn in place. The upper section of the sheet was left parted to expose the features to view and remained so until the time came to secure the lid. In this way the body was put into the coffin wearing just a shift and bonnet. It was not only a neat way to "finish" the interior but also gave the dressed corpse the appearance of being in bed.
>
> (Litten, 1991 79)

A winding sheet differs from side sheets in that the former is a more extensive piece of cloth which extends under the body and is held in position by the weight of the corpse, rather than being nailed in place along the lower edge.

Clothing the body

> We went to see a dead man laid out, and it would be a pity not to give you an account of his state. He was dressed in a long flannel shirt edged with lace, with flounces all down the middle, and five or six more on either side of the chest, all sewn with wool, in accordance with an Act of Parliament which forbids the use of linen or cotton for this purpose. The sleeves end in cuffs. This shirt is put over the head and does not surround the body in one piece but is fastened at the back. The body is laid out in a coffin resting on a bed of bran or sawdust. The face is shaved, the head dressed in a bonnet, fastened at the top, and a cravat and gloves of flannel are worn. When the lady is laid out, it is in a mob-cap and an embroidered headband, and neatly pressed flounces. The shirt comes right down over the feet, and threaded through with woollen thread at the ankles, with tassels at the end.
>
> (Dezallier-D'Argenville 1731, quoted by Litten 1991, 57)

Around 1700 grave-clothes changed (probably as a result of the influence of the soft-goods trade in ready-made items), adopting a form that was to last for the next 75 years. The winding sheet with top and bottom knots, of the 16th and 17th centuries, was replaced by an open backed, long sleeved shroud with draw strings at wrist and neck. The use of a cap or bonnet had been in vogue since the 1630s (Litten 1991, 76). Gloves, mittens, stockings and occasionally slippers were sometimes added to these basic items.

In the 18th and early 19th centuries the majority of commercially produced grave clothes were manufactured in London, but by the 1830s, with the increase in the use of cotton, Lancashire became increasingly important as a production centre (Litten 1991, 81). Not all grave clothes were bought ready-made but were often made by women of the household. This is attested by the Workwoman's Guide of 1838, which gives instructions for making caps and shrouds (Anon, The Workwoman's Guide 1832; reprinted Bloomfield Books, New York, 1975; quoted by Litten 1991, 82). With the repeal of the *Act for Burial in Wool* in 1815 a variety of textiles, including alpaca, calico, cambric, cashmere, etamine, flannel, holland, linen, muslin, poplin, satin, serge, and silk, for shrouds and winding sheets, with calico, cambric, flannel, linen, silk, and swansdown, for linings became available. (Litten 1991, 81).

5 Preservation of textiles and sample bias

Textiles decay differentially according to the burial environment, and there are always sample biases due to differential preservation among archaeological textiles (Edwards, 1974; Jakes & Sibley, 1983; Sibley & Jakes, 1984). The large quantity of material recovered from this excavation is largely a result of the relatively short time between deposition, and the difference between burial in vaults and in the soil. Nevertheless there is sample bias, which is witnessed not only by survival and presumed loss, but also by the relative condition of both the textiles and their constituent fibres. The majority of material is wool, cotton, and silk, with a minority of bast fibres, probably linen. The proteinic fibres, wool, hair, and silk, are usually the most robust in archaeological environments, although they will decay if the conditions are right (Jakes & Sibley, 1983; Sibley & Jakes, 1984). The cellulose-based fibres, linen and cotton, are easily degraded in the presence of moisture and air over a relatively short timescale (Janaway, 1989a). Cellulose is most susceptible to acids, while wool degradation is promoted by alkalis (Jakes & Sibley, 1983; Sibley & Jakes, 1984).

The first sign of degradation in wool fibres is shown by loss of cuticular scales and fibrillation. The silk fibres were generally the least degraded of those found on the site, exhibiting only slight morphological change, although having modified mechanical properties. Of the five examples of bast fibres found on the site, two were from fully anaerobic and three from very desiccated conditions. The low frequency of bast fibres may also be because they do not seem to have been used for 'funerary' textiles; four of the five samples come from 'normal' garments. Where bast fibres did survive they were in good condition, suggesting that conditions were such that they will either survive well or be totally lost. The condition of the cotton fibres varied considerably from those which retained many morphological features, to those which were swollen and fibrillated.

Inside a coffin, textile decay is unavoidably linked to the biochemical breakdown of the cadaver (Janaway, 1989b). This is demonstrated by the relatively poor condition of the woollen textiles from inside compared with those of the outer coffin coverings. The rate and nature of cadaveric breakdown is governed by a large number of factors (Polson *et al* 1985) (see Chapter 7). Important amongst these are the interval between death and the coffin being sealed, whether the seal was originally gas-tight, and the speed and scale of coffin breakdown by the time of excavation. It is self-evident that coffin construction plays an important part in this (Mant, 1989).

In most cases lead coffins probably had good gas-tight seals when first interred. However, the lead breaks down in time for two principal reasons: first, downward pressure from the practice of stacking a large number of coffins on top of one another results in a bowing of the side walls which exerts strain on the seams, causing them to come apart; and second, highly destructive lead corrosion precipitated by acid gases caused by air pollution (especially sulphur dioxide from burnt fossil fuels, which reached a peak in 19th century London) and from acetic acid released from wood in close proximity (Clark & Longhurst 1961; Evans 1961). This was demonstrated in the Christ Church crypt, where many of the lead coffins were no longer gas-tight when excavated, although the contents sometimes had a greater moisture content than adjacent all-wooden coffins (eg 2070, 2179, 2184, 2207, 2243, 2244, 2291, 2327, 2519, 2609). Some coffins retained a great deal of liquid, and had sufficiently low rates of gas exchange with the outside to retain anaerobic conditions until their opening by the excavators (eg 2188, 2263, 2373, 2407, 2575, 2634, 2680, 2749, 2764, 2811, 2822, 2916, 2949, 2955). In extreme cases anaerobic conditions can lead to a very high degree of soft tissue preservation, as in the case of 2188 (Fig 6.2). The inhibiting effect of lead ions on micro-organisms in this context is poorly understood. This high level of soft tissue and textile preservation can also occur over longer timescales as shown by the medieval body from St Bee's Priory, Cumbria (Tapp & O'Sullivan 1982; Glover 1990).

With wooden coffins, whether single- or double-shelled, the degree of sealing would depend on the quality of the timber, how quickly the boards warped, and the degree of insect attack. The quality of the joints in the first place would determine how quickly the seals broke down (Mant 1989). The use of pitch on coffin interiors would have had the advantage of providing an elastic sealing medium.

Putrefaction, which commences immediately after death, is a result of both bacterial action and degradation by the body's own enzymes (Polson *et al* 1985, 20). Aerobic organisms deplete the tissues of oxygen and, although their numbers decrease as the available oxygen is diminished, they create conditions which are more suitable for the more destructive anaerobic organisms that ultimately come from the intestinal canal. Most fungi found on decomposing remains are aerobic, and consequently their growth is restricted to the surface of the cadaver and associated textiles. In a well-sealed coffin, fungal growth will eventually become inhibited by the formation of a reductive atmosphere (Garland & Janaway 1989, 27).

Winged adult insects may lay eggs on the body, and hatch into larvae which feed before pupating; thus infestation must take place prior to burial or after the coffin has started to break down. Textiles can be damaged by larvae that are feeding on the body burrowing through them, as well as those that are deliberately feeding on the textiles themselves (Evans 1963, 5, 50, 51).

Fat is initially hydrolysed by the body's own enzymes, a process which proceeds slowly and soon diminishes. More effective than these intrinsic lipases are the enzymes produced by bacteria. Water is necessary for both the intrinsic and bacterial enzymes to work, though there is usually sufficient in the fat tissue itself. This may not be the case if the body dries out quickly. If the process continues, the neutral fat is totally converted to fatty acids, which are deposited in its place. If no further chemical changes take place these fatty acids remain as adipocere (Polson *et al* 1985, 21). The presence of adipose tissue implies at least locally anaerobic conditions. Fatty acids, the products of hydrolysis, will be quickly oxidized into aldehydes and ketones in the presence of oxygen, which is often absent in a sealed coffin. Protein is also broken down by enzyme action, the rate being determined by the amount of moisture, bacterial action, and temperature. Moisture favours decay, and proteolysis is slowed by cooling and increased by warming. As the proteins are broken down, the tissues become increasingly liquid, and it is this stage that has probably the greatest implications for the degradation of textile materials in the coffins. The precise chemical composition of this liquor at different stages of decomposition is poorly understood. Where this has been measured, it has proved to be alkaline, no doubt due to ammonia-based compounds from protein breakdown (Garland & Janaway 1989). An alkaline solution would promote the degradation of wool, as indicated by the degree of fibrillation and scale loss on a number of fibres from inside coffins. The degradation of cellulose-based textiles such as cotton and linen can be promoted by organic acids from wood.

As soft tissue starts to break down gases are formed which include carbon dioxide, carbon monoxide, hydrogen sulphide, ammonia, and methane. It

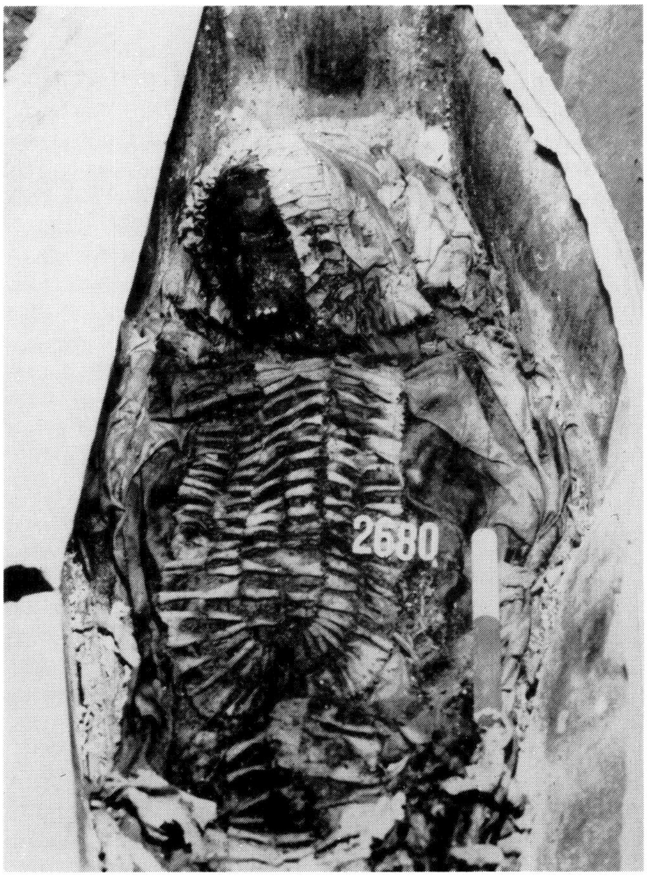

Figure 6.4 2680 body as excavated (skeletonized despite lead coffin) showing cap and shroud

is the formation of these gases, together with the ingestion of oxygen by aerobic micro-organisms, that converts the body tissues and the immediate burial environment (unless oxygen is free to diffuse it) to reducing conditions. The skeleton collapses when the ligaments have sufficiently decayed; in some instances the textiles were in a good condition over a semi-articulated skeleton (eg 2680 Fig 6.4).

In a rapidly drying environment, warm and with a good air flow, the tissues will desiccate and the degradation processes will halt, leading to natural mummification (eg 2273, 2274, 2324, 2325, 2329, 2331, 2333, 2334, 2351; Figs 6.1, 6.5; Polson *et al* 1985, 26–7). In particular, a dry substrate will adversely effect the enzyme systems of micro-organisms, especially fungi, which will result not only in the preservation of the cadaver but also of the associated textiles. Both, however, will be prone to insect attack (2325; Fig 6.5).

The site produced a relatively large number of burials with poor textile preservation, while a small number of either anaerobic lead coffins or

Figure 6.5 (below) 2325 body as excavated (naturally mummified) showing textiles pinned around head and under chin

Table 6.1 Preservation of fibre types in lead coffins

This table documents the number of identified textile samples recovered from the interior of LEAD COFFINS. The burial environment at time of excavation has been broken down into six categories.

Dry Skeletalised

Burial no	Cotton	Wool	Bast	Silk	Comments
2070		1			
2179		1			sawdust packing
2184		1			sawdust packing mould spots on textiles
2211		1		1	sawdust packing
2243			2		
2244		2			sawdust packing
2291	1				
2519			2		
2327		3			sawdust packing
2609			1	1	

Damp Skeletalised

Burial no	Cotton	Wool	Bast	Silk	Comments
2259		5			sawdust packing

Wet Skeletalised

Burial no	Cotton	Wool	Bast	Silk	Comments
2203				1	considerable dark brown wet residue
2255		3			
2307		1			
2917	2				

Wet, Extensive Soft Tissue, Anaerobic

Burial no	Cotton	Wool	Bast	Silk	Comments
2373	4		2		
2407	6				
2575	5				
2634		5			
2680		1			
2749		2			
2764		2			
2811		4			
2916	4		2		
2949		3			
2955		7			

Wet, Soft Tissue Present, Anaerobic

Burial no	Cotton	Wool	Bast	Silk	Comments
2263		1			

Dry, Extensive Soft Tissue, Anaerobic

Burial no	Cotton	Wool	Bast	Silk	Comments
2188		1			

very desiccated burials provided the most complete 'sets' of textiles. All the mummified bodies came from the north parochial vault. The bulk of identified textile samples from within coffins came from 72 lead and 30 wooden coffins. Of the latter, 9 contained mummified bodies.

The representation of different textile materials recovered from lead coffins is documented in Table 6.1. The burial environment at the time of excavation was classified into one of 4 categories, based on the nature of the body's soft tissue preservation and the moisture level. The textile samples used in this data exclude outer coffin coverings (CING). While the sample size is small a number of tentative generalisations may be made. Firstly the combination of damp conditions, resulting from extensive soft tissue decay and the presence of sawdust, would probably adversely affect the survival of cellulosic textile fibres such as cotton and linen. In the anaerobic lead coffins 26% of burials had cotton textiles, 34% woollen, 2% bast (linen) and 50% silk. Cotton was found far less frequently in lead coffins that were no longer anaerobic.

Table 6.2 presents the equivalent data for wooden coffins. Here there was a higher representation of wool and silk from the burials without mummified tissue, and a higher incidence of cotton associated with mummified bodies. However, as the presence of cotton is related to prohibition of its use prior to 1815 (see below) this may not entirely be a factor of differential preservation but also relates to the date and status of burials from the North Parochial vault.

The overall date range for the anaerobic lead coffins with textiles (Table 6.3) is 115 years. These were the coffins of 5 adult males, 7 adult females, and 2 infants. Eight of the adults were aged over 60 at death, while 11 were over 50 years old. Unfortunately, similar data did not survive for the mummified bodies because of the low preservation of the iron breastplates in the area from which most of these were recovered. It is likely that an overall interpretation of the textile finds may well be biased by an over-reliance on well-preserved material from one location in the crypt and from the most expensive burial facility (ie a lead coffin).

Different conditions can exist within the confines of a coffin; for instance the frill and top of the lining will be above the liquefied remains; whereas, the mattress, the back of garments, and the bottom of the winding sheets etc will be below. (This of course assumes that the coffin remained the correct way up!) The backs of a number of garments were considerably weakened compared to their fronts. The presence of sawdust in the coffins would promote the degradation of cotton and linen due to the liberation of organic acids. This would however only affect textiles with which it had direct contact (Clark & Longhurst 1961).

The silk facings etc from 2042, 2058, and 2203 were probably attached to woollen garments that have completely decayed. In addition to the differential loss of parts of garments, and the loss of some items completely, some textile remains were impossible to recover due to their degraded state or the state of the body. Well-preserved samples were more readily recovered. The frequency for the site as a whole, of for instance CING, LING, MATT, may be over-represented because their bulk assists recognition and sampling even in a highly degraded state.

6 Textiles from Christ Church: the evidence

It is difficult to make generalizations regarding cloth types and coffin types, since the best textile evidence comes from triple-shell coffins, which may be presumed to have the most complex textile sets (as they represent 'deluxe' burials), the problem especially being differential preservation between these and single-shell coffins.

The coffin exterior

Consists of: coffin outer covering (CING)

Cloth outer coverings for coffins were introduced in the 17th century. The majority of the outer coffin coverings were plain woven wool, often with a raised nap, nailed or stuck to the coffin sides before the decorative metalwork and round-headed studs were added. The fibres of the outer coffin covers were generally in the best condition probably because this class of textiles was not in contact with body decomposition products and wool fibres are generally robust in non-alkaline environments. The main agents of decay are insects and fungal mycelia, but the action of these was generally inhibited in areas adjacent to or covered by the coffin metalwork. A further reason for the large quantity of these fabrics is probably because of the ease with which samples could be recovered by the excavator.

This is the best represented category of textile finds from the site and is too numerous to list individually. A total of 144 CING samples were recovered from 222 burials which included textiles. In some burials, eg 2009, 2072 etc, CING samples were the only textile category preserved/recovered. Most samples are plain woven, woollen cloth. There are occasional exceptions, for instance 2142:01:US is a wool twill. Silk velvet outer coffin coverings were rare in the late 18th and early 19th centuries, and although silk velvet weaving took place in Spitalfields, none of these covers was found in the excavations. This may, however, be a result of differential preservation.

The coffin interior

Consists of: inner coffin lining, side sheets, frill, mattress and pillow

These textiles formed the finish for the interior of the coffin, or the interior of the inner most shell of double and triple shell interments.

The textiles of coffin interiors consist of a lining sheet (stuck to the sides and sometimes the base) or

Table 6.2 Preservation of fibre types in wooden coffins

This table documents the number of identified textile samples recovered from the interior of WOODEN COFFINS. The burial environment at time of excavation has been broken down into six categories.

Damp, Skeletalised

Burial no	Cotton	Wool	Bast	Silk	Comments
2013				1	
2015			1	1	sawdust packing
2017		1		1	
2020		2		1	
2023		1			
2034				1	
2035		1			
2037				1	
2038		1			sawdust packing
2039		1			sawdust packing
2042				1	
2045				1	
2059		1			
2060		1			sawdust packing not coffin but wooden box with three burials
2140		1			
2177		1			sawdust packing

Dry, Skeletalised

Burial no	Cotton	Wool	Bast	Silk	Comments
2047				2	
2058				4	sawdust packing
2077		1			sawdust packing
2082		1			sawdust packing
2087				1	
2091		1			sawdust packing
2098		1			
2119		1		1	sawdust packing
2134		1			sawdust packing
2137		1			
2146		1			sawdust packing
2152				1	straw/hay packing
2158				1	sawdust packing
2164		1			sawdust packing
2167		1			sawdust packing
2169		1			sawdust packing
2171					sawdust packing
2174		1		1	
2176	1			1	
2181		1			sawdust packing
2208		1			
2213				1	
2219				1	
2221				1	
2224	1				
2228				1	sawdust packing
2230				1	
2231				4	
2233				1	sawdust packing
2235	1				
2256	1				
2257	1				

Burial no	Cotton	Wool	Bast	Silk	Comments
2258	4				
2260	1				sawdust packing
2262	2		1		
2264	1			1	
2268		1			sawdust packing
2272				2	shroud and cap too degraded to sample
2275	1			1	
2297				1	sawdust packing
2347				1	sawdust packing
2357	1				
2369	1	1			sawdust packing
2396				1	
2422				2	
2446				1	

Dry, Extensive Soft Tissue

Burial no	Cotton	Wool	Bast	Silk	Comments
2101		1			sawdust packing
2267		3			(ligaments plus internal organs preserved)
2311		1		1	

Dry, Extensive Soft Tissue, Mummified

Burial no	Cotton	Wool	Bast	Silk	Comments
2273	1				
2274	5	1	1		sawdust packing
2324	3				
2325	4				
2329	9	1	2		
2331	8				sawdust packing
2333	2				
2334	3				
2351	2				

side sheets fixed at the top, but hanging loose, curtain-like, at the sides. Frills are used to conceal the lining decoratively (Fig 6.6) and/or the wooden bead around the top of the coffin. The function of the top bead is to support the inner lid.

Mattresses and pillows were placed at the base of a coffin underneath the body (Fig 6.7). Fewer pillows than mattresses were recovered (although when in a very decayed state discrimination between them could be difficult). These categories represent the largest group of materials, but have had the least work done on them. They consist of cloth covers, often decorated with frills or rosettes, which are then stuffed with a number of materials including wool, hair, and straw. The stuffing material, especially hay, is of environmental interest. There are plant and animal remains from field, workshop, and associated with the body's decomposition, all of which would repay future investigation.

Coffin bases were filled with sawdust, wood shavings, etc, which are found on top of the mattresses, and were probably added after the body had been viewed, immediately prior to the coffin being sealed. Their purpose was twofold: firstly to absorb liquid from the body's decomposition, and secondly to stop the body rolling and bumping while the coffin was being moved.

Inner coffin lining (LING)

number of wool lining = 18
number of cotton lining = 10
number of silk lining = 3
(burials:2045,2213,1446)
total number of lining samples = 51
Due to time constraints not all LING samples were analyzed

This category is too numerous for individual listing. There was some confusion in recording and it was not always possible to make distinctions between inner coffin lining (directly adhering to sides or base of coffin), side sheets (inner coffin lining attached only at top edge and free hanging) and occasionally winding sheets [WRAP] (which are used to

Table 6.3 Dates of death, age and sex of anaerobic interments with textile samples

Date	Sex	Age
1731	M	69
1739	F	73
1758	M	50
1763	F	
1767	F	90
1779	F	59
1789	M	51
1793	F	63
1797	F	45
1798	M	60
1823	F	68
1824	F	65
1825	M	
1847	M	64

wrap the body and are not fixed at all to the coffin structure). Where there has been any ambiguity in recording, the sample has been assigned to the LING category rather than SDSH or WRAP. The linings were plain cloth, without pinked or scalloped edges, and without punched decoration. Side linings can have all these or can be plain.

The majority of coffin linings were either of plain woven cotton or wool, in the same fabric as mattress covers, frills or shrouds. There were three burials that possibly had silk linings; the fragment from 2045 is a very degraded woven textile now totally unravelled (possibly a silk/linen mix) that could have been part of garment rather than a lining. The samples from 2213 and 2446 are less ambiguous fine silks in a satin weave.

A number of lining samples have pitch adhering to them, evidence that the lining was used to cover the coffin interior after it had been sealed in pitch.

Frills (FRIL) or rufflings were used to conceal the top edge of coffin linings and side sheets; they are often attached to the wooden bead used to support the inner coffin lid. They usually have pinked/scalloped edges with punched decoration. Tucks are usually held with tacking stitches and secured by iron tacks. The fabrics used were plain woven, either wool or cotton.

Cotton frills:
2235:01, 2256:02, 2257:02, 2260:02, 2262:02, 2273:01, 2274:03, 2324:03, 2329:03, 2331:05, 2333:03, 2334:04, 2373:01, 2407:01

Woollen Frills:
2327:03, 2634:01

Mattress (MATT) usually shaped to cover the whole of the coffin base. Cloth covered, with various stuffing materials: wool, feather, or hay. The covers are usually plain woven of either wool or cotton. See Table 6.4.

It might be assumed that sawdust coffin packing was not necessary when a mattress was used, there are however six examples where both a mattress and sawdust were recorded as being present, but this may be the result of confusion in recording. Time allowed only a small number of MATT cover samples to be analyzed and the stuffing material was not identified.

Pillow (PILW) used to support the head. Cloth covered, with various stuffing materials : wool, feather, or hay. The covers are usually plain woven of either wool or cotton and match the other coffin lining fabrics. Most burials with pillows do not have mattresses, the exceptions being 2243 and 2373. Only a small number of PILW cover samples were analyzed and the stuffing materials were not identified.

Table 6.4 Mattress (MATT)

number	cover	fill	coffin type	coffin packing
2019:01	-	-	wood	sawdust
2020:01	wool	-	wood	none
2023:01	wool	-	wood	none
2029:03	-	-	wood	sawdust
2039:02	-	-	wood	sawdust
2077:04	-	-	wood	sawdust
2153:01	-	-	wood	sawdust
2154:02	-	hay	wood	none
2243:01	-	-	lead	none
2244:01	wool	-	lead	sawdust
2327:02	wool	-	lead	sawdust
2331:04	cotton	-	wood	sawdust
2373:02	cotton	-	lead	none
2609:01	wool	-	lead	none
2811:04	wool	-	lead	none

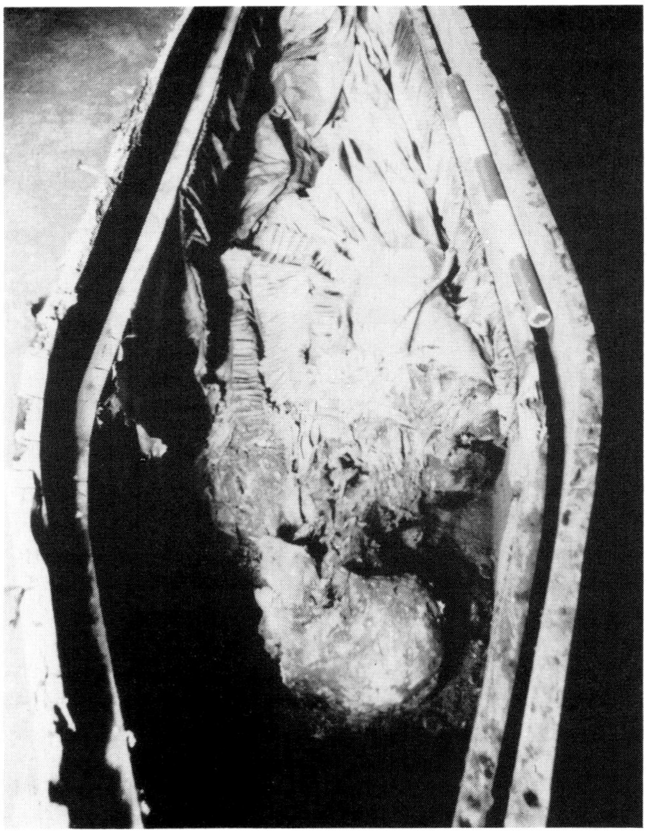

Figure 6.6 2324 body as excavated (naturally mummified) showing frill on inner coffin, shroud and shroud sheet

Pillow Samples:
2052:02, 2059:01, 2060:01, 2066:02, 2142:02, 2142:03, 2151:01 (fill hay), 2155:03, 2160:01, 2164:01, 2171:01, 2173:02, 2177:01 (wool cover), 2210:03, 2211:02 (silk and wool cover), 2215:03, 2222:01, 2224:03, 2225:01, 2230:02, 2231:04, 2243:02, 2259:03, 2260:03, 2264:04, 2267:02 (cotton cover), 2287:01, 2313:02, 2373:03 (cotton cover) 2407:02 (cotton cover), 2575:01 (cotton cover), 2634:02 (wool cover), 2811:05 (wool cover), 2916:03, 2955:01 (wool cover)

Textiles associated with laying out the body

The Christ Church burials provide evidence for broad trends relating to laying-out the body. The variations recorded appear to result more from differences in individual practice, as today, rather than any general differences between treatment according to age at death, sex, or changes through time.

There is only one recorded example of a pad being bound to the loins (2325, MF 6.36). In modern practice this is to absorb the decomposition fluids that are evacuated from the anus soon after death. The remaining loin coverings are 'modesty clothes' used to cover the body's sexual organs. If a procedure similar to modern practice were used, the person laying out the body, would have stripped and washed the body. Lower garments and sheets in contact with the body at death are often soiled and these are discarded and the body covered. In a modern undertaking practice it is recommended that clean underclothes are put on the body, to cover nakedness. No obvious underclothes were found at Spitalfields, but the plain cloth on 2351 probably fulfilled this function. Once the body was placed in coffin other textiles would of course cover this.

A number of ties were used to stop the body from moving out of position in the coffin. In particular the legs were tied together, either at the ankles or by the big toes, and the arms were sometimes tied to the sides of the body. In modern practice medical bandages are used, but at Christ Church, the ties were sometimes torn-up strips of other materials (eg 2058:02:LP) or more commonly they were plain

Figure 6.7 2634 pillow

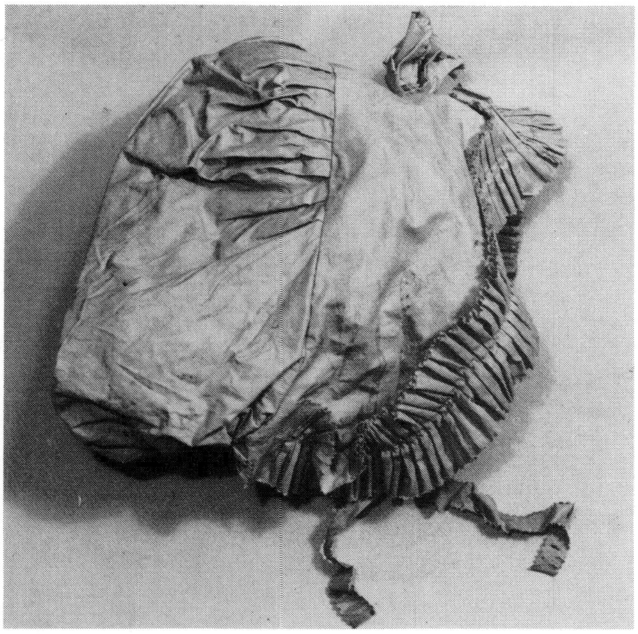

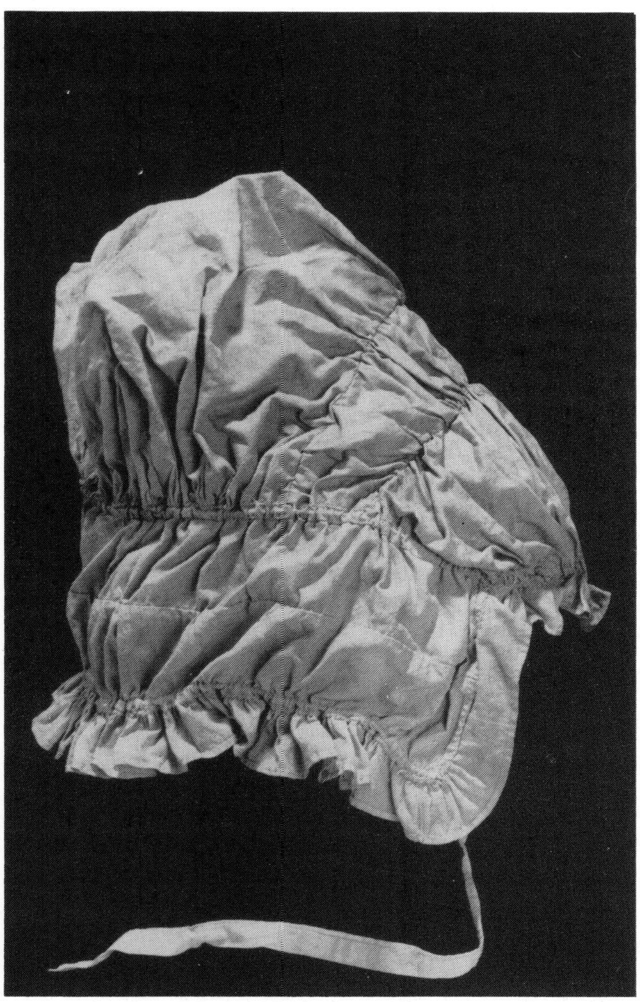

Figure 6.8 (above) 2575 woman's cap of type not made for funeral

Figure 6.9 (below) 2407 cap made for funeral

commercially produced silk ribbons (eg 2231:07:NE) In one instance, a brocaded ribbon was used for this purpose (2176;01:US; MF 6.15). It is clear that the 18th and 19th century undertaker had plain silk ribbons available in the same way as a modern undertaker has bandages. The exceptional use of fancy ribbons or strips of cloth would seem to represent the use of whatever was to hand. In the catalogue all items categorized as TTIE were used in this manner, while RIBB can be used for ties, but may also be part of fastenings of caps and other head coverings etc. The function of trussing a body in this manner is to keep it in a neat, seemly position while being viewed, and to prevent the limbs banging against the coffin sides while the coffin is being moved.

The problem of the jaw dropping open is addressed in a number of ways. It can be tied up by silk ribbons (eg 2044; MF 6.5), or a special cloth could be pinned to the cap, called a 'jaw cloth' (eg 2325 Fig 6.5). Sometimes this item has the same decoration as the shroud and linings. One example recorded in an excavator's notes has a piece of cloth like a triangular bandage, wrapped twice around the jaw and head, and tied under the chin (2351). A cap was then placed on top of this, which is not shown in the drawing. In some instances the under-chin ties of caps were probably sufficient to keep the jaw closed.

There were some examples of small squares of cloth being placed over the face, called 'face cloths'; these were, like the jaw cloths, of similar material and decoration as the rest of the funeral cloth sets. One odd example (2916:04:LS) was found on the throat. It is of different material to the rest of the linings etc. and is not decorated; given its size, and the lack of pins or other fastenings, it is unlikely to be a support for the jaw, and is probably a face cloth that slipped after the coffin was closed.

There is little evidence for coins being used to close eyes. Certainly the bodies were not generally buried with coins on the eyes. The exception is 2420.

The preparation involved in the use of jaw cloths, face cloths, and general presentation of the body, supports the view that during the 18th and 19th centuries it was normal for the coffin to be open prior to the funeral.

The body coverings can be divided into two groups:

(a) Material which comes as part of the burial: frills, linings, shrouds etc. (Figs 6.4, 6.6, 6.9, 6.11–13)
(b) The deceased's own clothes, wigs, stockings, etc. (Fig 6.1, 6.9, 6.18, 6.19)

In practice, provided the remains are not too fragmentary, it is possible to distinguish between the two categories (eg 2407:03:EP, a cap made for the funeral, Fig 6.9, in contrast to the cap 2575:03:EP, Fig 6.8). The main criteria used to make this distinction were based on the range of cloth types in any one burial, the presence/absence of punched decoration and the degree of finish (hems, stitch types etc). 'Burial clothes', made of similar cloth to the linings and frills etc, often had pinked, scalloped edges with punched decoration; they lacked hems, and very coarse tacking stitches were used in a generally crude construction. They often had decorative strips of gathered material attached to the front faces. Of course, with some items, such as the backless shrouds, their funerary nature was self-evident. Where items of normal clothing were included, the quality of construction and needlework, even of the poorest examples, is infinitely better than that of the funerary textiles. The latter do not have buttons or buttonholes, being fastened when necessary with either ties or copper alloy pins, and they lack linings, facing, pockets, or lace. The decoration on normal garments was not plain-punched, as this would not stand up to wear without having the edges oversewn, as in broderie anglaise. Some garments, such as 2331:09:NP, show evidence of repair (MF 6.43). Even the best quality of the funerary textiles, such as the cap 2917 (Fig 6.10) and winding sheet 2407 (Fig 6.11), are still readily distinguishable from the normal clothing from burials such as 2916 (MF 6.47).

Many of the bodies were wrapped in a winding sheet (WRAP) (Fig 6.11), usually having scalloped/pinked edges and borders of punched decoration. These were large sheets, big enough to wrap the body, often consisting of two loom widths of material stitched together at the selvedge in order to give the necessary size. The punched designs can range from simple to complex. In a number of cases there is evidence that although the body was completely wrapped the head was exposed (2324), although not in the case of 2331.

Under the winding sheet there could either be a shroud or clothes (or sometimes both). All the clothes sufficiently preserved were upper body garments, with an extra length of cloth to cover the upper legs, such as with 2354 or 2916. The only lower body garment, except stockings, were the drawers described, but not recovered, from 2519. The garments recovered included shifts, shirts, chemises, jackets and a waistcoat. These are described individually below.

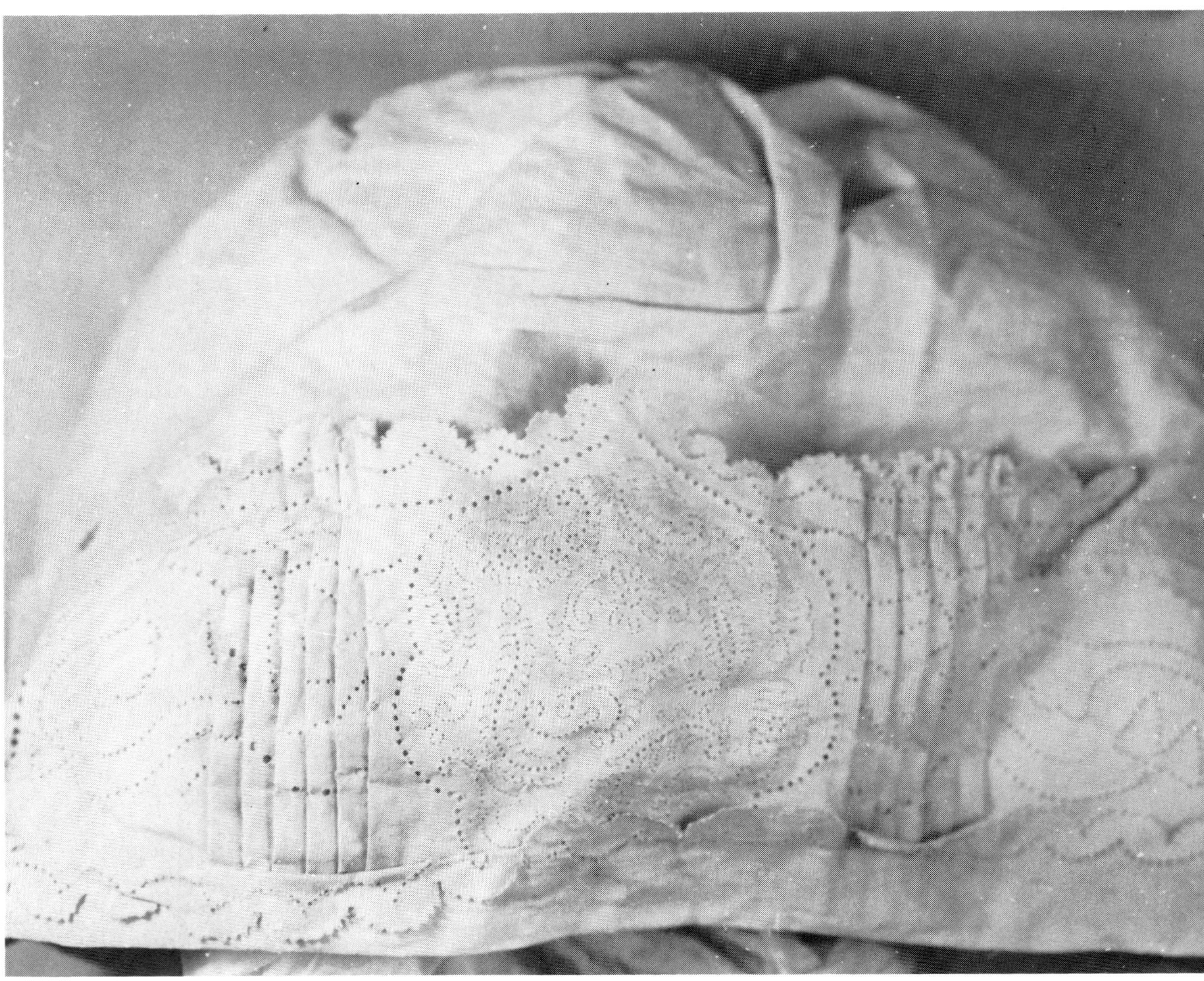

Figure 6.10 2917 front of cap made for funeral

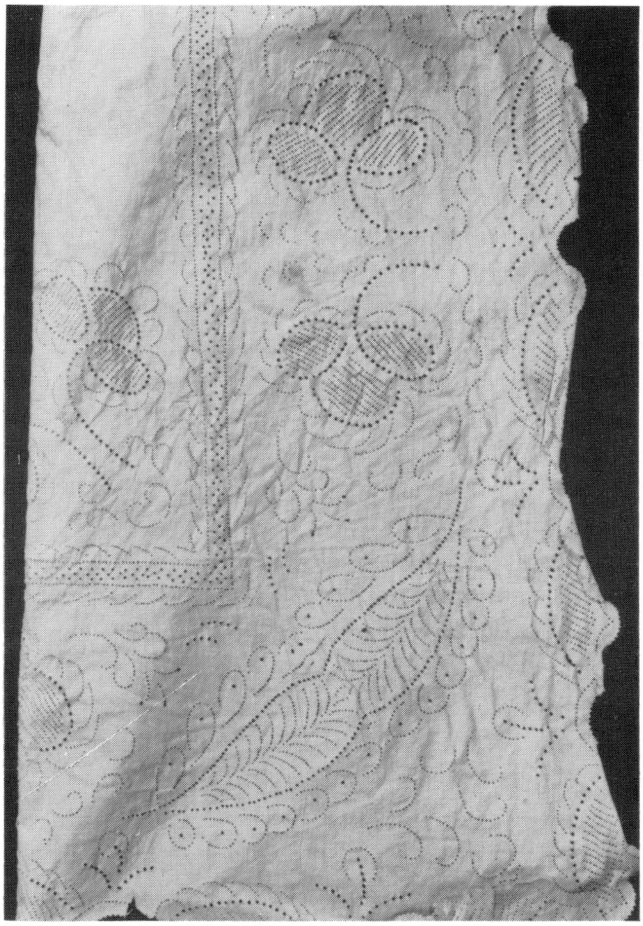

Figure 6.11 2407 detail of punched decoration on cotton winding sheet

Figure 6.12 2917 shroud

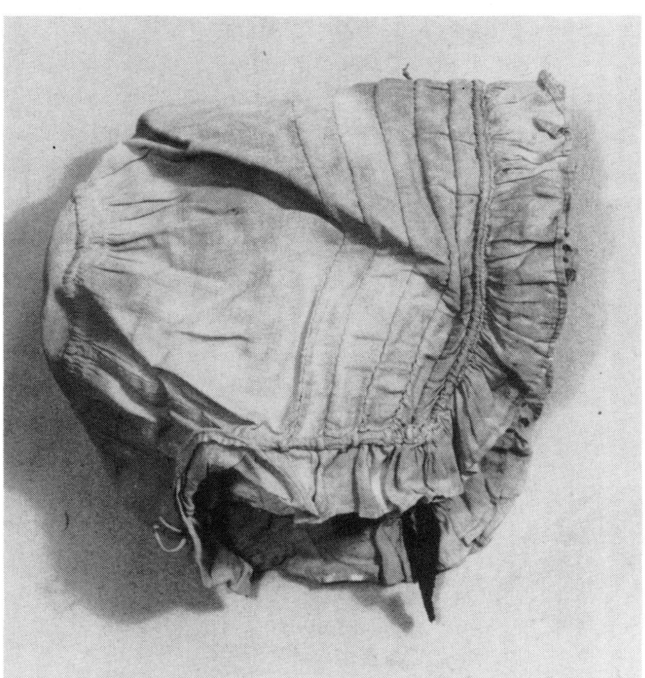

Figure 6.14 2373 baby's cap

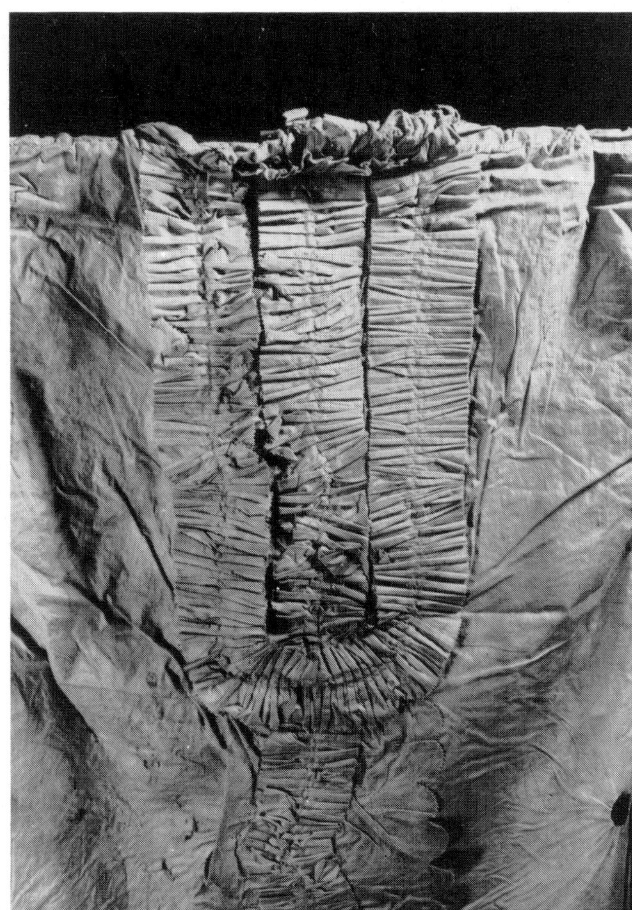

Figure 6.13 2917 detail of shroud front

Shrouds are backless, but with sleeves, and cover the body from the neck to the feet. They are decorated with gathered frills on the region of the chest, and often with punched decoration (Figs 6.4, 6.12, 6.13).

There is no evidence that all the bodies had caps, or other head coverings, although many may have been lost due to differential degradation within the coffin. The caps range from those made especially for the funeral (eg Figs 6.9, 6.10) to the occasional item of 'normal dress' (eg Figs 6.8, 6.14).

Two individuals were buried with men's knitted silk gloves (2219, 2221; MF 6.18–19), while from 2955 (MF6.49) came a pair of mittens made especially for the funeral. Seven pairs of stockings were recovered; in all cases, where sufficiently preserved, they reached to above the knee (MF 6.40, 6.50). All were knitted and made from wool, silk, or cotton (see below). More of the bodies were probably buried in stockings, but examples of these may have been lost due to degradation and difficulties in the recovery of material from legs. In one burial there was a footless sock which had been rolled up to provide an extra support for one side of the head. This had probably been removed from the body during laying-out, and was simply to hand; an example of the purely fortuitous inclusion of the deceased's clothing in the coffin.

There are no indications as to the relationship between the use of 'grave clothes' as opposed to 'normal clothes'. The body could have clothes with a winding sheet, or a winding sheet and shroud, or winding sheet and shroud with clothes underneath. The nature of the shroud is such that it covers the whole body from neck to feet, so no clothing underneath would be visible. Although it could be argued that as a thriving second-hand clothes market flourished in London in the 18th and 19th centuries, the inclusion of clothes could be taken as a sign of wealth, this does not seem to be the case from the Spitalfields evidence. On the other hand, a shroud would not be required if the deceased was buried in their own clothes. One possibility is that in the event of an individual dying from what was thought to be an infectious disease there was reluctance on the part of the person laying out the body to disturb the clothing. Again the wishes of the deceased or their family to be buried in a favourite garment such as a wedding dress, ballgown or uniform could have influenced the decision. One burial (2329) had a winding sheet, silk stockings, and an elaborate satin and lace-trimmed garment, while another (2331) had a shift and jacket, coarsely made and repaired. There is evidence for both 18th and 19th century garments from the site.

The evidence for children comes from three burials (2188, 2250, 2373). Burial 2188 (Fig 6.2) was dressed in a woollen cap, shroud and winding sheet, made especially for the funeral, and similar to those used for adults. There is no information for 2250. Burial 2373 had a pillow, cap, and gown that had not been made specifically for the funeral (Fig 6.14).

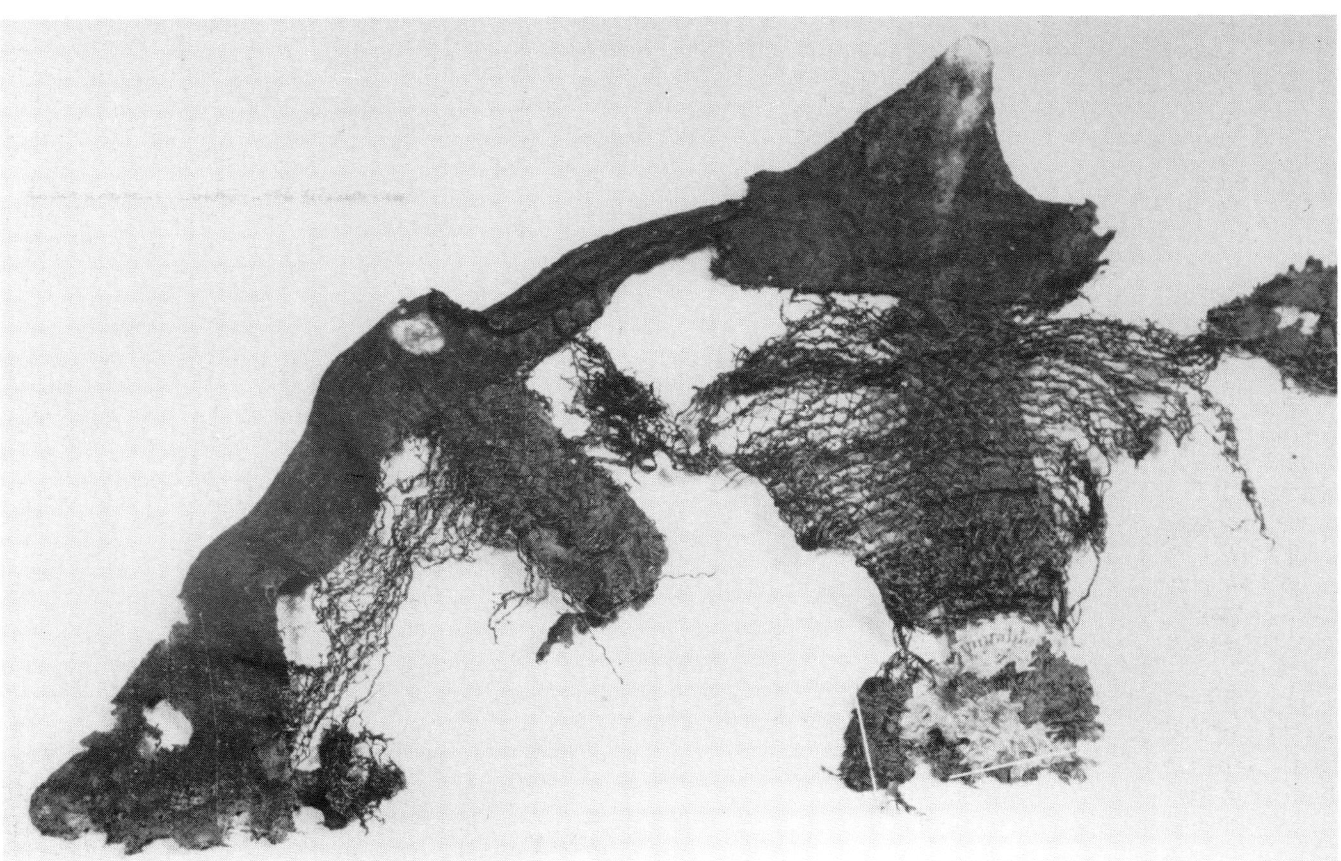

Figure 6.15 2017 wig base

Figure 6.16 2013 pig tail

Textiles associated with the body : the head

Textiles associated with the head consisted of caps, wigs, as well as face cloths, and cloths used to support the jaw. According to Litten (1991, p76) the cap or bonnet had been a feature of grave clothes since the 1630's. *The Workwoman's Guide* of 1838 gives details of how to make head gear for the grave.

> 'CAP. If the usual cap is not put on, the following is made for a man: it is of flannel, cut exactly like an infant's foundling cap. A quilling of punched flannel is put round the face, and a band of it is laid on behind, and across the top of the head, strings of the same, are also sewed in. CAP FOR A WOMAN. This is of flannel: the round part is platted up to the front, and a quilling of the bordering put on, a band of the same laid on at the back, and strings.'

Wig or hair piece (WIGG)
Four burials had remains of wigs or hair pieces:

> 2017:03 a wig base on brown silk plain weave, silk net with a white silk maker's label which is only partially legible. The hair is missing (Fig 6.15).
> 2243:03 a hair piece with long dark hair affixed to small fabric base from a burial in 1749 of a woman aged 60.
> 2013:01 and 2037:01 are pigtails or queues, ie human hair wrapped in a silk ribbon. Since in each case both ends of hair have been cut, these may be examples of a 'false queue' (Figs 6.16, 6.17).

Head covering (HCOV)
These were mainly caps, either made specifically for the funeral (same material and decoration as shroud, winding sheet, linings etc), or items of normal dress.

Caps (made for the funeral) (Table 6.5)

Caps (normal dress) (Table 6.6)
Other:

> 2158:01 fragment twined silk, found in head area from the burial of woman who died in 1759 aged 49 years.
> 2158:02 fragment of silk yarn, found in head area (probably part of 2158:01)
> 2228:03 a ribbon, 1130 mm long and 65mm wide, found on head
> A knitted jersey hat was found, but not retained (Fig 6.3).

Face cloth (FACL): rectangular piece of cloth used to cover the face. Often with punched decoration, and pinked/scalloped edges, usually made of the same material as the shroud, lining, and frill.
Five facecloths were recorded:

> 2259:04 (wool) 2575:02 (cotton) 2680:01 (wool) 2822:01, 2955:02 (wool)

'Jaw cloth'(JWCL): strip of fabric either tied or pinned in place to prevent the jaw dropping open.

Two wool jaw cloths were recorded. 2255 was accompanied by a cap, while the excavator's notes do not mention a head covering in the case of 2749.

2347:01 is of silk, with seamed edges and is fixed under the chin, and worn with elaborate cap (non-funerary) (MF 6.44).

Neck cloth (NECK): rectangular piece of cloth placed on neck, with no evidence of being fixed as with JWCL.

There is only one example 2916:04. This is unusual in being made of bast fibre, probably linen. According to the excavator's notes the rectangle of cloth was found 'under the chin'. The individual was wearing a cap, but there is no evidence of this cloth being pinned to it to secure the jaw.
CRAV One fragment of wool cloth 2955:05 with three folds along the long axis, scalloped, pinked, and punched decoration, and laid lengthwise down chest. It had no obvious function.

Figure 6.17 2037 pig tail

Table 6.5 Caps made for the funeral

Textile no	Material	Sex	Age at death	Date of death
2243:04	wool	F	60	1749
2255:03	wool	M	36	1768
2259:05	wool	F	48	1754
2263:02	wool	F	73	1739
2264:02	cotton	M	07	1832
2274:04	cotton	–	–	–
2407:03	cotton	F	65	1824
2916:05	cotton	M	64	1847
2917:01	cotton	M	69	1836
2949:01	wool	F	45	1797

Textiles associated with dress: the body

The degree of differential preservation and recovery

It is impossible to make generalisations about how the body was covered because of the combination of differential preservation of the textile materials and differences in recovery and recording by the excavators.

This is illustrated by a number of burials in which only a head, leg or hand covering was recovered/recorded:

wig only: 2013, 2037
cap only: 2017, 2152, 2184, 2228, 2231, 2263, 2264, 2347, 2369
stocking only: 2179
glove only: 2219, 2221

Sixteen bodies were dressed in one or more garments while eleven were covered by only a shroud. Two had both garments and a shroud. Six burials with shrouds had caps, while six burials in garments also wore caps.

Shrouds

Shrouds are backless garments, with sleeves, usually with ruffled decoration on the front. They are crudely sewn with long tacking stitches and often a number of the edges are cut with pinking shears.

Shroud only (Table 6.7)

There is a strong correlation between the presence of shrouds and the use of a lead coffin; nine out of ten burials with shrouds were in lead coffins; the exception is 2060, which was in a wooden box containing three burials rather than a coffin.

In six out of ten burials the shroud was accompanied by a cap. The shroud and cap are made from the same material in all cases. Both male and female burials had caps and shrouds. The earliest date for a burial with a shroud was 1732, the latest 1836.

Shroud + body garments (excludes gloves,socks etc)

In two instances the body was dressed in both garments and a shroud.

An 82 year old woman buried in 1843 (2267) was wearing a fragmentary cotton garment with a frill at the cuff and neck openings, fastened with cartwheel buttons. It is not possible to tell if this was part of a night gown or some other garment. Over this was a cotton shroud. Burial 2331 was interred with a cotton shroud covering cotton stockings, a cotton, long-sleeved, open-fronted jacket, fastened with a tie at the neck, worn over a shift.

Garments but no shroud

Fifteen burials produced evidence of garments but no shrouds. In a number of instances the textile re-

Table 6.6 Caps (normal dress)

Textile no.	Material	Sex	Age at death	Date of death	Description
2329:11	cotton	–	–	–	complex weave fabric, edged with lace and fastened with a silk ribbon
2351:01	cotton	–	–	–	decorated with broderie anglaise and trimmed with lace
2369:01	cotton	F	55	1831	with lace trim
2373:04		M	–	1825	baby's cap, linen cloth with silk underchin ribbon
2575:03	cotton	F	68	1823	

mains recovered from that burial are very fragmentary eg 2020, 2203 etc.

2020 : fragment of brocaded silk presumably from garment.

2042 : silk-satin and twill facing, and pocket flap and lining from men's coat or waistcoat, with button holes.

2045 : very degraded fragment of silk.

2058 :a number of silk fabric samples recovered from the burial, including silk gauze, part of facing of garment (records lack sufficient information for detailed reconstruction).

2203 : silk facing of collar edge or coat tail, three button holes from man's coat. Man, 58, buried in 1795.

2262 : small fragments of linen garment, with fragment of cuff with cartwheel button, recovered by excavator but with insufficient detail for further reconstruction. Evidence of very degraded cotton stocking on legs.

2274 : linen shirt fastened at front with alternating cartwheel and mother-of-pearl buttons, with cotton cap and cotton stocking.

2325 : fragments of cotton shirt cuff band, three cartwheel buttons.

2329 : five samples from complex, high quality women's garment; made of linen, with lace trim to cuffs and neck. Tied with ribbon, terminating in lace-trimmed lappets, under bust. Cuffs drawn with silk satin ribbons. Worn with silk stockings and a cotton/silk cap.

2351: long-sleeved, cotton, open-fronted jacket, fastened with cartwheel buttons at cuffs and neck and worn with cotton cap.
2373: baby's cotton gown, open-fronted, fastened with tie at neck, long sleeves with triangular gusset under arms worn with cap of linen and silk. Male infant buried 1825

2519: open-fronted wool jacket with long sleeves, fastened with four buttons, worn with wool garments on lower body described by the excavator as 'leggings', but not recovered. Man, 52, buried in 1837.

2575: cotton open-fronted shirt, button on collar only, cuffs drawn with cotton ribbon, triangular gusset under sleeve. worn with cotton cap and cotton stocking. Woman, 68, buried in 1823.

2609: rectangular, hemmed silk gauze (diagonal tear repaired with overstitch); not hemmed, held in place over body with three silk ribbons. Woman, 78, buried in 1791.

2916: linen shirt with long sleeves worn under an open fronted cotton jacket worn with cotton cap, cotton stockings. Man, 64, buried in 1847.

Body covering (unable to distinguish between garment and shroud)

There were thirteen burials where the body was covered by textiles but it was impossible to ascertain whether these samples represent parts of garments or shrouds. The materials recorded are:

silk: 2078, 2158
wool: 2174, 2208, 2244, 2307
cotton: 2176, 2224, 2258, 2311
too degraded to identify: 2070, 2159
leather: 2172 (fragment of fine leather, stitched, with copper alloy pin inserted)

Winding sheet (WRAP) a rectangular piece of cloth on which the body was laid, which is then folded over the body. Usually with pinked/scalloped edges and border of punched decoration. Material can be either wool or cotton. Sometimes termed coffin sheet.

2274:02:NP cotton cloth
2324:04:NP cotton cloth
2329:02:NP cotton cloth
2331:03:NP cotton cloth
2407:04:EP cotton cloth

Table 6.7 Shroud only

Burial	Shroud material	with cap?	Sex	Age at Death	Date of Death
2181	wool		M	64	1744
2255	wool	wool cap	M	36	1768
2259	wool	wool cap	F	48	1754
2327	wool		F	23	1793
2407	cotton	cotton cap	F	65	1824
2749	wool		M	51	1789
2811	wool		M	50	1758
2917	cotton	cotton cap	M	69	1836
2949	wool	wool cap	F	45	1797
2955	wool	wool cap with wool gloves	M	69	1732

2060 is not a burial in a single coffin but a wooden box with three burials, of which two had shrouds

2634:03:PV wool cloth
2764:01:PV wool cloth
2916:06:LS cotton cloth
2949:03:LS wool cloth
2955:06:VE wool cloth

Textiles associated with the body: the hands

Glove or mitten (GLOV)

Three burials were excavated where the body was wearing gloves:

2219:02 pair of man's knitted silk gloves, with short cuff.
2221:02 pair of man's knitted silk gloves, with short cuff.
2955:07 pair of wool mittens, made for the funeral, with scalloped edges and punched decoration.

Textiles associated with the body: the legs

Stocking (STOC)

All the samples recovered, whether of wool, silk or cotton, are knitted with a seam up the back of the leg. All that were sufficiently preserved reached above the knee.

Cotton: 2258:06, 2274:05 2331:07, 2575:05
Silk: 2329:10
Wool: 2179:01, 2634:05

Slipper (SLIP)

2634:06:PV Only one burial produced any type of footwear. This was of a pair of slippers, made of a heavily fulled woollen cloth, in three pieces; sole, heel, and upper. The slippers were of different sizes and were worn over woollen socks. They were found on a 60 year old man, buried in 1798, who possibly suffered from gout.

Leg covering (LCOV)

2916:07:LS piece of cotton cloth laid across upper legs, in similar manner to that found on 2331.

7 Examples of well-preserved groups of textiles

A number of burials had well-preserved groups of textiles. Some of these are discussed below, in order to indicate the range of data that can be retrieved from a single coffin. Where known, details of sex, age and date of death are given.

2259

Female, 48 years old, 1754

This burial was in a lead coffin, no longer fully sealed, and the body largely skeletonized.

The coffin lining was wrapped around a wooden bead running round the top of the inside walls, over which a frill was attached. The body was wrapped in a winding sheet, ie, placed in the centre of the sheet and each long side was folded over the body. The edges of the sheet were cut into scallops with pinking shears and the border had punched decoration (MF 6.28). A 'face cloth' covered most of the face and was of the same material as the winding sheet. The body was sketched by the excavator (MF 6.25). It was covered with a shroud and wore a cap of the same material (MF 6.27); both had the same punched decoration and pinked/scalloped edges as the rest of the textiles. Below the body was a full-length mattress which had a finer plain weave fabric on the upper surface rather than the lower. The coffin had been packed with sawdust. The fabric of all the items, except the mattress, was a very similar, plain woven woollen cloth.

2274

(No biographical details surviving on coffin plate)

This body was in a single-shell wooden coffin from the north parochial vault and was naturally mummified.

The interior of the coffin sides was covered by free-hanging cotton 'side sheets' which were suspended from wooden beading. The wooden beading and the top of the side sheet were covered by a decorative cotton frill. The side sheet had complex punched decoration, including flowers and a winged cherub, while there were two types of punch-decorated designs used on the frill. The body was dressed in a linen shirt (MF 6.31) with the front fastened with alternating 'mother of pearl' and 'cartwheel' buttons. The legs and feet were in cotton stockings with the feet tied together by a silk ribbon. A cotton funeral cap was worn on the head.

2324

(No biographical details surviving on coffin plate)

The body was in a double-shell wooden coffin from the north parochial vault and was naturally mummified (Fig 6.6).

The interior of the coffin was in a good state of preservation, with side lining, frill, winding sheet, and shroud more or less intact, especially from the area of the shoulders down to the feet. A cotton frill was attached to the top of the inner edge of the inner coffin. Behind it an extra strip of cloth was attached with tacking stitches. The cotton lining was held in place with nails. The frill has pinked, scalloped edges and fine punched decoration, with holes of two different sizes. The winding sheet, wrapped around the body (MF 6.32) had punched decoration along the borders similar to that of the frill. It was wrapped squarely around the body with the top edge turned back to reveal the head. Under this was a shroud, with punched decoration at the bottom and cuffs. The cuffs were constricted by the use of ties made from the same fabric, cut with pinking shears. The front of the shroud was decorated with applied, gathered strips of cloth. Under this was

Figure 6.18 2329 detail of material, and lace trim

worn a smock (MF 6.33) with a smocked front and full sleeves gathered into wide cuff bands (with buttons). Unfortunately it was not possible to recover this garment.

2325

(No biographical details surviving on coffin plate)

The body was in a double-shell wooden coffin from the north parochial vault and was naturally mummified (Fig 6.5).

The body was covered in a shroud with both pleats and rosettes on the front; underneath was a loin cloth, which was not sampled (MF 6.36) . At the foot of the coffin was a cloth fragment, probably part of a garment, hemmed in a triangle, with a monogram in one corner (IG in blue thread) (MF 6.35).

2329

(No biographical details surviving on coffin plate)

The body was in single-shell wooden coffin from the north parochial vault and was naturally mummified (Figs 6.1, 6.18, 619). The cotton frill and lining were both hanging from wooden beading. Unusually, the frill was of an open-weave gauze, with pleats held in place by a strip of cloth (also cotton) sewn together with a tacking stitch. At the base of the coffin, loose fibres, thought to be the stuffing from a mattress, were discovered.

The upper body was dressed in a complex, high quality garment, which had disintegrated below the pelvis. The main fabric of the garment was bast fibre, probably linen, the top of the sleeves being let into the arm hole with a series of fine tucks (sample 2329:04:NP). The long sleeves were finished at the wrist with cotton lace and drawn together with a silk satin ribbon tied in a bow (sample 2329:05:NP). The backing of the cuff band was made from two grades of plain woven cotton cloth. The garment was tied below the bust with a strip of plain woven cotton cloth, which ended in two lappets, edged with cotton lace (MF 6.37, 6.38)

The legs were covered by knitted silk stockings, which reached to just above the knee. Fig 6.19

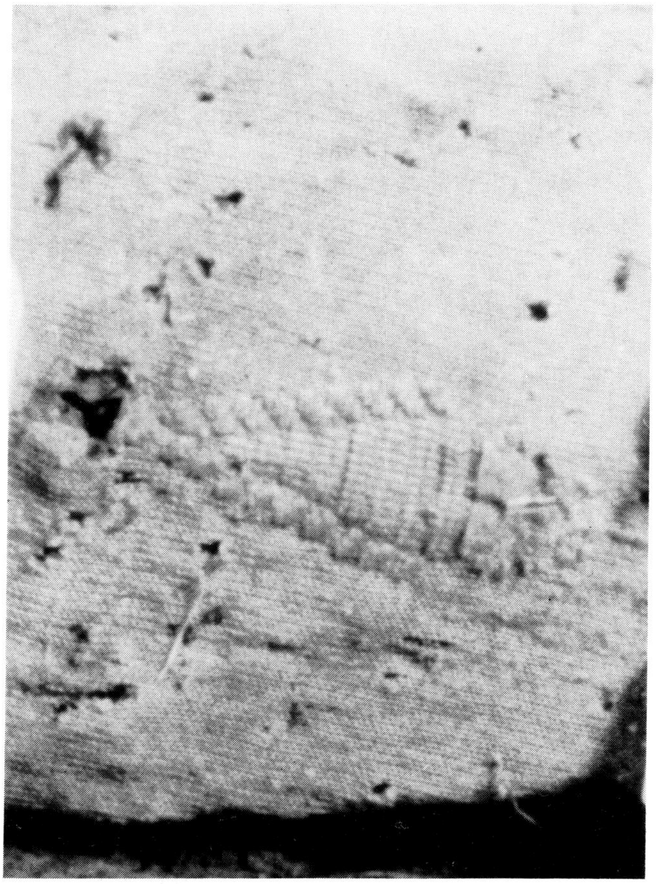

Figure 6.19 2329 detail of clock on stocking

shows a decorative motif, which could not be examined in detail because it was only possible to detach a small sample from the stocking top for analysis.

A small fragment of a cotton cap, with a complex decorative weave and trimmed with lace, was recovered from the head area. Like the body garment, it is clearly a 'normal' item of dress, not made for the funeral; it was fastened under the chin with a silk ribbon.

The body was wrapped in a winding sheet, which had scalloped edges cut with pinking shears and a textured surface, similar to 2325:02:NP (MF 6.34).

The quality of the textiles from this interment suggests that to be buried in clothes and wrapped in a winding sheet, rather than having a shroud and winding sheet, could be a question of choice, rather than economic necessity. Possibly this body was buried in a dress with some special significance to the deceased, or her immediate relatives, such as a wedding dress.

2331

(No biographical details surviving on coffin plate)

The body was in a double-shell wooden coffin from the north parochial vault and was naturally mummified (Fig 6.20).

The coffin lining, which was in good condition, was in the form of a 'side sheet'; ie it was only attached along the top edge of the coffin and finished at the bottom of each side. The top edge was covered with a frill, which had pointed, pinked edges, and punched hole design.

The base of the coffin was covered by a mattress, approx 1.3 m in length, shaped to fit the coffin, with eight rosettes on the upper side. There would originally have been more rosettes, but part of the head end of the mattress has been lost. The mattress is finished with a frill with a scalloped, pinked edge. The mattress was 10–20 mm thick, covered with cotton plain woven cloth, and stuffed with hay. The fabric covering the top is of finer quality than on the bottom.

The body was wrapped in a winding sheet, made of a plain woven cotton fabric with pinked, scalloped edges. It has an unusual punched decoration, consisting of a single row of holes made with a star-shaped punch. The sheet was tied around the feet, and placed around the head and shoulders in a 'hood' arrangement.

Under the winding sheet was a cotton shroud with a fine punched design. Beneath the shroud was a short jacket, with long sleeves, a gathered collar, and tied at the neck (MF 6.41/2). The jacket was made of poor quality plain woven cotton cloth and the left sleeve was made of two pieces of fabric, joined at the selvedge to make up the length.

The legs were covered with knitted cotton stockings, complete except for the toes. They came over the knee, and the tops were turned down (MF 6.40). The tops were finished with two rows in red yarn, and the initials CC were embroidered in cross stitch. There was an opening in the seam to accommodate a drawstring, though none was used.

Underneath the jacket and over the stocking tops was a plain woven cotton shift (MF 6.43). There is a patch of darning near the left shoulder, and the initials CL (possibly CI) embroidered just below the neck.

2339

(No biographical details surviving on coffin plate)

The body was in a double-shell wooden coffin from the north parochial vault and was naturally mummified.

The body was dressed in a knitted jersey-style hat, with a neckerchief around the neck, a long shirt, and stockings (Fig 6.3). It was wrapped in a winding sheet, which had pinked, scalloped edges. There was a frill attached to wooden beading around the top of the coffin.

2351

(No biographical details surviving on coffin plate)

The body was in a double-shell wooden coffin from the north parochial vault and was naturally mummified.

The coffin was lined with cotton cloth, stiffened with glue and stuck to the sides of the coffin; around the top was a frill of cotton gauze. The body was wrapped in a winding sheet, with pinked, scalloped

Figure 6.20 2331 body as excavated (skeletalised) showing frill on inner coffin, shroud sheet and clothing

edges and punched decoration. There was no shroud, but the corpse was dressed in a jacket, stockings, and the upper legs covered in a plain cloth. The cotton jacket was gathered at the neck using fine tucks, which were hidden from view by a gauze (frill) collar. The front opening was closed with two copper alloy pins and a 'cartwheel' button at the neck. The sleeves were inserted into the cuff-band with fine tucks. The cuffs were finished with a ribbed woven cotton frill and fastened with two cartwheel buttons [cuff saved, 2351:02:NP].

The ankles were tied using a strip of coarse woven cloth. The jaw was tied shut with a chin strap, formed from a square of cloth, rolled up in the manner of a triangular bandage, and tied under the chin. Over this was a bonnet, not made for the funeral, held in place with a copper alloy pin. The main fabric was a plain woven cotton, with alternating thick and thin bands of weft to give a ribbed effect, decorated with tucks, broderie anglaise, a gauze frill and lace trim. It was tied with a ribbon.

2373

Male, 11 months old, 1825

Double-shelled facility, lead coffin providing anaerobic conditions, with the soft tissue preserved. When the coffin was opened it was full of dark brown liquid, containing a black sediment. There were small traces of adipocere, with the skeleton inside a ball of textiles. There was a strong smell of ammonia.

There were two side sheets, damaged where they were hanging into the liquid. There were no fixing points visible on the linings, suggesting that they were probably glued. The frills were pleated, with scalloped/pinked edges and punched decoration. The pleats were held in place with a tacking stitch, and originally nailed in place. It seems likely that the frill was nailed into the top edge of the outer wooden coffin and hung down into the inside of the lead coffin when this was open. The frill was then removed and dropped inside when the lead coffin was sealed, resulting in the loose tacks found in the bottom and the position in which the frill was found.

The body rested on a mattress and pillow, both with plain woven cotton covers and stuffed with feathers. It was dressed in a long-sleeved cotton gown, with triangular gussets under the sleeves, open at the front and tied at the neck. On the head was a cap (Fig 6.14), with a silk ribbon fastening made up of two cotton fabrics, the finest being used for the edging frill. It had not been made specially for the funeral.

2407

Female, 65 years old, 1824

Triple-shelled facility, with lead coffin providing anaerobic conditions, and the soft tissue preserved. The lower half of the coffin was filled with dark brown liquid. All the textiles below waist level were impossible to sample due to the extent of degradation and thick, smelly adipocere.

The lead coffin was lined with 'side sheets' and had iron tacks along the top edge for fixing. The cotton frill was found lying loose in the bottom with its tacks in place. It had a pinked/scalloped edge and punched decoration. Like the frill from 2373, this was probably detached before the lead coffin was sealed.

On the base of the coffin was a mattress which was not sampled as it was heavily soaked in decomposition products. The head was on a pillow, which was covered in a plain woven cotton cloth with rosettes stitched to the upper surface.

The body was wrapped in a cotton, punch-decorated winding sheet, the top half of which was recovered almost complete. This covered the whole body including the face. It went under the body but over the mattress. Underneath the winding sheet was a shroud, of which only the top half survives. This was made of a fine plain woven cotton, with gathered applied strips of cloth and rosettes. Decorative frills had pinked/scalloped edges and punched decoration. On the head was a cotton cap (Fig 6.8) which was made specifically for the funeral. This had a decorative strip, similar to those on the shroud, and a strip of cloth, with pinked edges, tied in a bow and stitched to the top.

2503

Female, less than 1 year old, 1826

The burial was in a triple-shell facility, including lead coffin, with complete wooden inner coffin. Conditions were anaerobic, although there was no liquid. The body was subject to a complete autopsy, but no textile samples were retained.

A frill, which was pleated, punched, and pinked around the top edge, was nailed along the top edge of the wooden inner coffin sides. Hanging below this were 'side sheet' linings. The coffin base had been lined by covering the base with sawdust and then nailing a cloth cover over this. The nails were arranged in a diamond pattern to give an effect like a buttoned sofa. Under the head was a very decayed pillow with a pinked/scalloped edged and punch-decorated frill.

The body was dressed in a gown, gathered at the wrists, with two drawstrings at the neck, one going under the arms, and the other round the neck.

As this garment was considerably larger than the body, the excess material had been carefully folded underneath or around the sides of the body. On the head were the remains of a cap, with green staining from the copper alloy pins that had been used to fasten it in place. The big toes had been tied together.

2519

Male, 52 years old, 1837

The burial was in a triple-shell facility, with a lead coffin; the contents were dry and the body fully skeletonized.

The upper body was dressed in an open-fronted, long-sleeved jacket, made of a heavily dressed plain woven woollen cloth. The open front had four buttonholes, faced with silk, which were fastened with 'mother of pearl' buttons. Similar buttons also fastened the cuffs. This was the only textile retained from the burial [2519:01:EP and 02:EP].

On the lower limbs were a pair of long, woollen, men's drawers; these may originally have been footed, although the foot end was decayed on excavation. They had a waistband of a finer worsted wool than the rest of the garment, which was described by the excavator as coarse. The waistband had a front opening, closed with two 'cartwheel' buttons. It was not possible to remove this garment.

2575

Female, 60 years old, 1823

The burial was in a triple-shell facility, with lead coffin and complete wooden inner coffin. The conditions were anaerobic, with the soft tissue in the form of adipocere going into a white and sticky sediment, with the skeleton exposed. The textiles were damp, but no liquid was present in the coffin.

No details were recorded of a lining to the coffin, and it is suspected that a number of items may have been present, such as a mattress, but were not sampled, and were also omitted from the context sheet. The head was resting on a cotton-covered pillow with hay stuffing.

The body was wearing a long-sleeved cotton chemise, with the cuffs drawn with a cotton ribbon and a frilled edge to the collar. The garment was open-fronted, with a button fastening on the collar and triangular gussets under the arms. The collar and cuff frills were made of a plain woven cotton which had been given a ribbed effect by using alternating bands of single and paired weft threads. On the legs were a pair of knitted cotton stockings.

The jaw was tied up with a cotton handkerchief (with hemmed edges and a repair patch of another fabric); this was folded on the diagonal and held in place with copper alloy pins. On the head was a cotton cap (Fig 6.9), not made specially for the funeral. This was tied underneath the chin.

2609

Female, 78 years old, 1791

The burial was in a triple-shell facility, with lead coffin, but was dry.

In the bottom of the coffin was a mattress with a plain woven woollen cover and wool stuffing. The upper surface was decorated with rosettes, which had punched decoration and pinked edges. The body seemed to have been wrapped in a square of very light silk gauze; this had a diagonal tear which had been repaired by oversewing (MF 6.48). It is difficult to ascertain whether this was a garment or merely a covering (probably the latter). It was edged or tied in place with three long silk ribbons. There was a pair of 'elasticated garters or ties' around each leg just above the knees; there is a possibility that these were stocking tops, but the excavator did not think so. They were not sampled. A ring was found on left side next to the finger bones.

This is an unusual burial in terms of the textiles recorded and recovered particularly as the main body cover was semi-transparent. There are no details as to how this was positioned, although a tuft of pubic hair was found adhering to it during conservation. It is unfortunate that it was not in a better state of preservation or recorded in more detail.

2634

Male, 60 years old, 1798

Triple-shell facility with sealed lead coffin and complete inner coffin. The cadaver exhibited extensive deposits of adipocere. The textiles on the corpse had almost completely decayed, with remains between the legs suggesting there had either been a shroud or some other garment in that region. The legs had possibly been bandaged, in addition to the wool stockings and slippers. The linings and frill of the coffin were in better condition than the rest of the textiles. There was a small amount of liquid in the base of the coffin.

The lining, extending down the sides and over the base of the coffin, was held on the base with nails and on the sides with glue. The body was covered by a shroud; fragments of a different fabric underneath the shroud suggested another garment. The wool slippers were of different sizes; this, along with the leg bandages, may suggest some deformity of the legs (gout?). The feet had been tied together.

The pillow had a woollen cover with a decorative edge and was stuffed with rags. There was no sign of a cap or cover on the head.

2916

Male, 64 years old, 1847

Triple-shell facility, with sealed lead coffin. There was a complete inner wooden coffin, with no liquid present, although the textiles were damp to the touch. The body's soft tissue was well preserved, the brown-coloured skin being intact.

The inner coffin was lined with a side sheet and frill. The body was wrapped in a white cotton winding sheet (Fig 6.11). This had a pinked/scalloped edge, and a broad border of fine punched decoration. It was drawn over the body, but not the face. The body was dressed in a linen shirt (MF 6.47), with a short neck opening, fastened at the collar and cuffs by buttons (missing when excavated). There are triangular gussets under the arms and the initials I K are embroidered in a red cross stitch. Over this was a short, open-fronted, long-sleeved jacket of cotton twill cloth (MF 6.47). This was fastened in front by four buttons, which were missing. A rather shapeless cotton cap, made especially for the funeral, was worn on the head. It had frills with scalloped/pinked edges and punched decoration, and an under-the-chin tie.

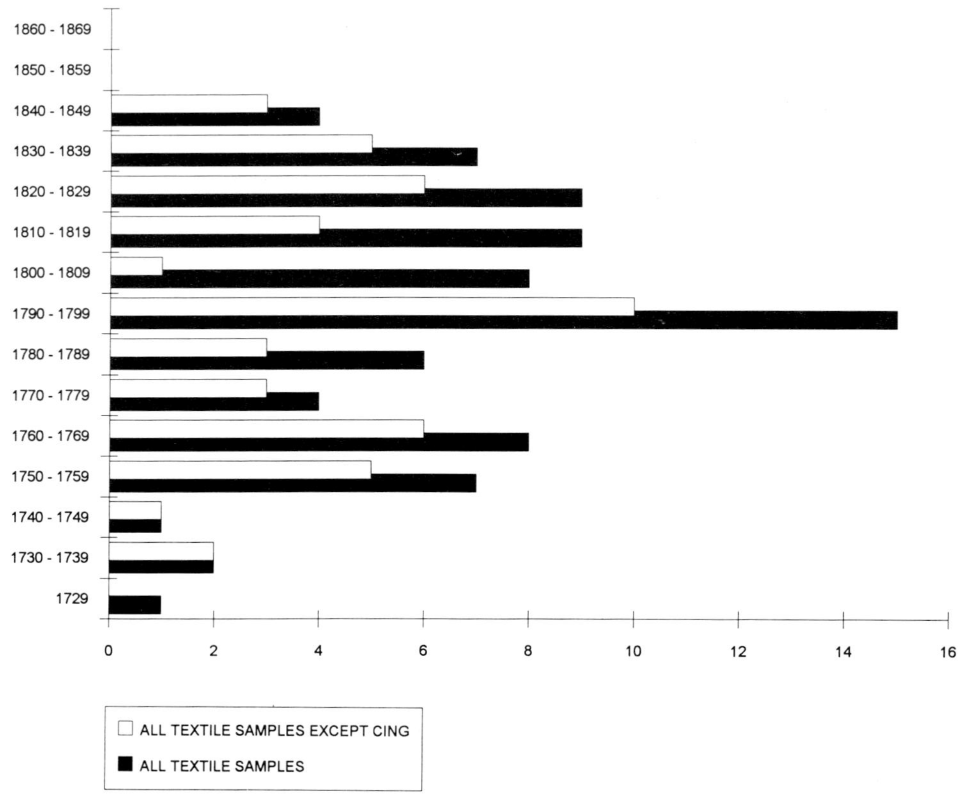

Figure 6.21 Frequency of textile samples from burials of known date

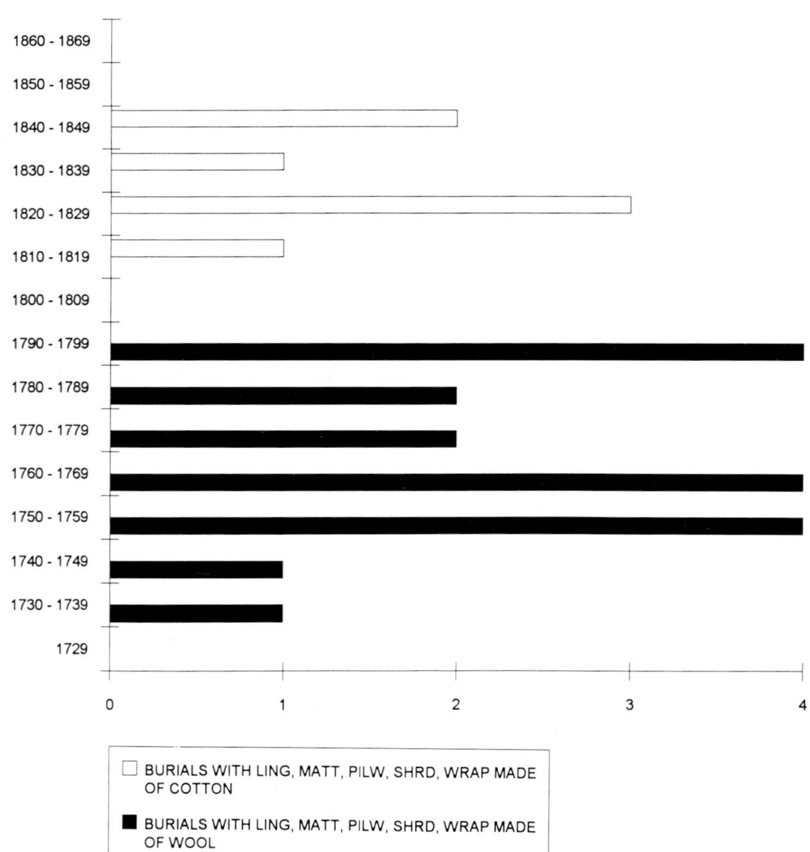

Figure 6.22 Frequency of woollen and cotton funerary textiles from burials of known date

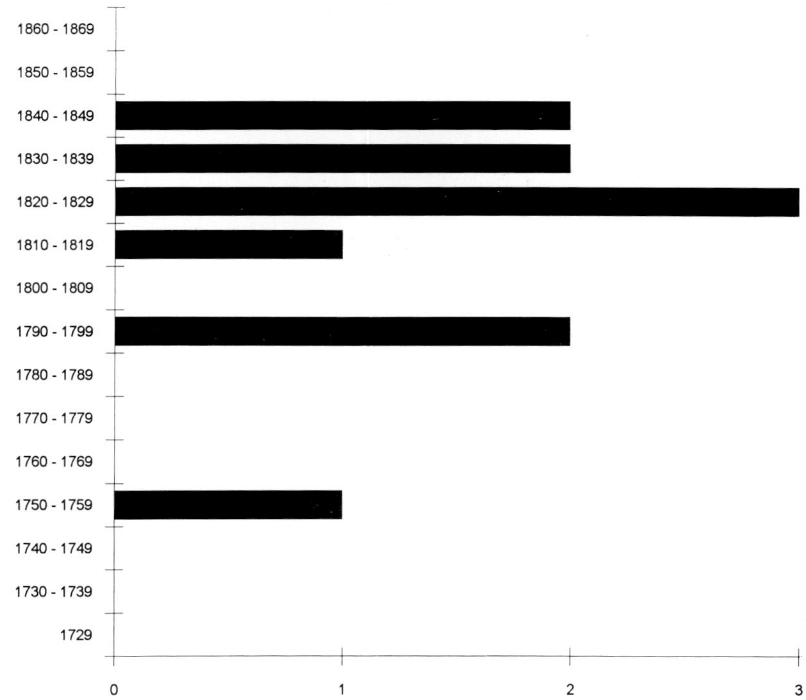

Figure 6.23 Frequency of non-funerary clothing (GARM, HCOV, GLOV) from burials of known date

A complete piece of cotton cloth, with no hems, but two selvedges (the other two edges had been torn), was laid over the upper legs and shins as a 'modesty cloth'. A small piece of linen cloth was folded over the neck.

2917

Male, 69 years old, 1836

Triple facility with lead coffin; there was no liquid, and although the textiles were damp to the touch, the cadaver had decayed to a skeleton. Only two items were recovered: a cotton shroud and a cotton cap. The shroud had very fine punched decoration on the bottom panel. The cap (Fig 6.10), made for the funeral, was of unique pattern for the site, being in the shape of a gentleman's smoking cap, with very fine punched decoration on the front. Originally there had been a cloth covering the face, but this was not recovered.

2955

Male, 69 years old,1732

Triple-shell facility with sealed lead coffin. All the textiles were a plain woven woollen cloth, with scalloped/pinked edges and punched decoration.

The inner coffin was lined with side sheets and a frill, which were found in place. A complete pillow was recovered, although there is no record of a mattress being found. However, in this case, this may be an omission of recording or differential preservation. The body was covered by both a shroud and winding sheet. The body was dressed in a cap and mittens (MF 6.49). All items were made for the burial.

8 Evidence for changes in burial practice through time

The burials at Christ Church, Spitalfields should enable a study to be made of the changes in the use of textiles in burials from 1729 to 1862. However, the example from the vault clearance of St Marylebone Parish Church of an octogenarian buried in clothing which would have been fashionable on someone 60 years his junior (Litten 1991, 73) points to the danger of basing the date of an the interment on the style of the material incorporated into it. Therefore, trends through time can only be examined using burials with secure dates from coffin plates. These data are presented in Figure 6.21, although it should be noted that the sample size for recovered and identified textile materials from inside the coffin is small (n=49). Figure 6.22 plots the frequency of burials with funerary textiles (LING, MATT, PILW, SHRD, WRAP) made of either cotton or wool. The distribution clearly indicates that these burials at Spitalfields were complying with the Act of the Burial in Wool. This was not repealed until 1815, when the first dated burial including cotton appears on the site. Figure 6.23 plots the frequency of use of non-funeral clothes. This is a very small sample, and it is probably unwise to generalise from these data, although as presented it seems that the bulk of these materials date from 19th century. From the excavations there are some silk facings from men's garments presumably made of wool (eg 2203 from 1795) which date from the 18th century.

7 Formation processes at Christ Church

1 The formation of mortuary deposits

> It seems surprising that there has been little explicit concern with the formation processes of mortuary deposits.
> (Chapman & Randsborg 1981, 11)

It is generally accepted that mortuary sites and settlement sites yield very different kinds of information about the past (O'Shea 1984, 24), and are generally investigated using equally incompatible techniques. Archaeological studies of the living (from settlement sites) and of the dead (from mortuary sites) of extinct societies have moved in separate directions since the 1960s. Whilst the former have adopted broad theoretical models from socio-economic and geographical scholars and the latter have tended to concentrate on largely material issues, it has been shown that analytical techniques developed in one discipline may be brought to bear on the other. Meggitt (1965), Saxe (1970), Goldstein (1981), and others have stressed that the application of spatial analysis to mortuary data may yield significant patterning in mortuary behaviour. More recently Chapman and Randsborg (1981) and O'Shea (1981) have demonstrated the need to apply models of natural and cultural processes outlined by Schiffer (1976) to account explicitly for the formation of mortuary deposits:

> The existing theory of mortuary differentiation has concentrated on statements which specify the relationship between the organization of a living society and its practices for the disposal of the dead. It fails to predict the additional relationship between these mortuary practices and their archaeological observation. . . . If there are regularities in these transformations, recognition of them could vastly enhance the accuracy and reliability of archaeological reconstruction of past social organisation.
> (O'Shea 1981, 40)

Although an understanding of formation processes is implicit in archaeological investigation, and increasingly well-covered in archaeological literature (Schiffer 1976; 1987; Wood & Johnson 1978; Kristiansen 1987), formation processes and mortuary studies have assiduously avoided mutual recognition until very recently (eg Boddington *et al* 1987). Exceptions to this general state of affairs have been provided by taphonomists (Gifford 1981; Shipman 1981; Mant 1987), who initially attempted to account for differential preservation in the fossil record, and who became increasingly concerned with understanding processes operating within archaeological environments, especially those acting upon human remains.

The present antipathy can be accounted for briefly as follows. First, formation processes literature has tended to adopt an artefactual approach. Schiffer (1987) treats mortuary deposits within the framework of cultural deposition, as a discard process. This approach concentrates on the attributes of artefacts recovered from mortuary deposits, and echoes the work of Kristiansen (1985) in attempting to account for the representativity of remains from prehistoric burial sites. The artefactual perspective on mortuary remains can indeed be traced directly back to the notable work of Worsaae (1843; 1849). Mortuary deposits, within this tradition, are treated as assemblages of material, rather than as areas of mortuary activity.

Second, mortuary deposits are unique because of the 'purposive and conscious elements' in their formation (O'Shea 1984, 24). The range of activities associated with the burial of human remains is vast and complex, as stressed in the work of Binford (1971) and others, and the processes which transform non-artefactual classes of evidence from mortuary sites, such as the human body itself, its disposal facility, and its spatial relationships with the rest of the mortuary environment, are as yet little understood — see especially the work of Mant (1987) on factors influencing the preservation and decay of human tissue. O'Shea (1984, 24) stresses the need for a separate theory of mortuary deposition.

However, attempts to account for the distinct nature of mortuary deposits and generate a set of theories expressly concerned with mortuary formation are as yet in their infancy. There are several contributory factors to this state. The application of well-tested experimental techniques to the modelling of mortuary data are beset with practical, ethical, and legal problems, and ethnographic observation of mortuary ritual has often been problematic because of the arcane nature of mortuary ritual in many societies. In addition, the opportunity to examine a mortuary population for which our knowledge of its living counterpart is anything like complete is rare.

There is, however, a context in which some of these problems may be lessened. While the depositional pathways which form mortuary deposits are unique, the processes operating on deposits after their deposition are in many cases similar to those which operate on other types of site. These post-depositional processes are increasingly well-documented and have been discussed fully by Schiffer (1987), Garland (1987), and Henderson (1987). The most important of these are the chemical, biological, and mechanical transformation of organic remains, the disturbance of deposits by flora and fauna including humans, and potentially large-scale transformation of the mortuary environment itself. A mortuary environment which reflects a population very similar to a living population, and in which post-depositional processes are easily identifiable, would be an ideal context from which to obtain data which might aid in the generation of some mortuary depositional models. Such a context exists at Christ Church, if we accept that demographic techniques are capable of determining the closeness of the Spitalfields population to our own.

2 Depositional context at Christ Church

At Christ Church deposits were formed in a discrete and identifiable space over a period of some 150 years, between the construction of the building after 1714 and the abandonment of the site around 1867 (see chapter 1.4). Post-depositional processes operated from 1714 until the end of excavations in 1986, a period of 272 years.

The spatial extent of archaeological examination was more or less exactly the same as that of mortuary deposition, though it did not include a crucial area in which mortuary activity was almost certainly taking place, since the greater part of the space beneath the church was remodelled during the 1960s without any form of archaeological recording. Thus the question of whether the excavated deposits represent something close to a complete sample of interments at the church assumes a critical importance (see chapter 1).

The potential for identifying post-depositional transformation is considerable. The vaults were abandoned and sealed around 1867, and remained isolated, with a few exceptions, from cultural disturbance until archaeological examination. Parts of the church, which apparently did not include those containing burials, were used for various activities, including shelter from air raids during the Blitz. Prior to 1984 the only identifiable cultural disturbance after abandonment was the introduction of electrical cables through some of the burial vaults during the 1960s. The impact of this disturbance appears to have been minimal: some disturbance of the upper surface of the dumps was evinced by footprints. During one such period of activity an empty cigarette packet (Players' Navy Cut) seems to have fallen through a borehole on the upper level (*c* 1984) into the Lemaistre\Pontardant vault, providing an interesting, but misleading, *terminus post quem* for the latest cultural deposition 'sealed' in the vault.

The extent of non-cultural transformation of the abandoned deposits can be reconstructed with some confidence (see section 5 below). The processes most easily identified were graviturbation, that is the compression of the deposits under gravity, and some downward movement of artefacts; faunal turbation; fungal and other biological decaying processes associated with wooden coffins and human tissue, and chemical decay of organic and inorganic materials, especially ferrous metals.

A discussion of the implications of this type of depositional context for the generation of a set of theories of mortuary deposition awaits further analysis, but is summarized by Adams & Reeve (1987). It is certain, however, that in archaeological terms the Christ Church data possess a high level of inference quality (IQ).

3 Identification and recording

An integrated method for the recording of formation processes as part of the strategy for archaeological excavation has not yet been developed, but there are signs of a willingness to accept the necessity of such a system in order to increase control on inferences made from archaeological data. The recording system at Christ Church was such that it was relatively easy to reconstruct formation processes after the event, and the photographic policy was geared towards the identification of processes. Nevertheless with the benefit of hindsight the excavators are aware that certain types of data were not recorded with sufficient detail to allow for the accurate reconstruction of several processes. These include the lack of on-site taphonomic analysis, with the result that the influence of micro-environmental factors on the decay of organic materials is poorly documented. In addition, an adequate sample of coffin wood for the purpose of determining decay resistance was not taken because of health and safety restrictions. For these reasons alone it is regrettable that a similar site is unlikely to be investigated in the foreseeable future (see Reeve & Adams 1986).

Despite these shortcomings a relatively full account of the formation processes at Christ Church is presented in the hope of controlling inferences made from the Christ Church data, and encouraging the identification of formation processes from other mortuary sites in the future.

4 Cultural formation processes at Christ Church

Cultural formation of the deposits within the vaults at Christ Church has been divided into four phases for discussion:

1 Primary deposition
2 Secondary deposition and maintenance

3 Abandonment
4 Excavation

For each phase there will be a general discussion of the nature of the processes operating, a description of the major factors affecting identification and recovery of processes, and some analysis of the inference potential of the processes inferred. Definitions follow those of Schiffer (1987) as consistently as possible, except where specifically stated in the text.

Pre-depositional processes would ideally be integrated with this section. Here, however, it seems more useful to follow Binford (1971) and others and include these processes under the section dealing with the undertaking industry and treatment of the body prior to disposal (see chapter 4). This is because the sources available for this type of analysis are very patchy, and more easily dealt with by established historical techniques. It must be stressed, however, that mortuary processes operating from the time of death to interment constitute an integral part of the formation of the mortuary deposit, and should be dealt with as such in future synthesis of the material. These processes, it must be remembered, play a highly significant part in the degree of preservation observed in human remains by the excavators (Henderson 1987).

Primary deposition

Primary deposition was taking place at the site from 1714 almost continuously until 1852, and then again for a short period around 1867. The formation of non-mortuary deposits, including the lowering of the floor, for which there is no direct archaeological evidence, the construction and destruction of internal walls, and the possible use of one vault as a boiler room, seems in general to have taken place either before or after the main period in which burials were taking place at the church. Other cultural deposits during the period 1729 to 1852, when all the mortuary deposits themselves seem to have been created, were either adjuncts to the process of burial, or accidental inclusions. Inscriptions, mostly carved on limestone slabs and attached to walls, were introduced either at the time of purchase of a private family vault, after an interment, or after the closure of one of the main parochial burial areas. Very few artefacts seem to have been discarded during this period of primary deposition. The overall sparsity of lost or discarded items generally implies either the operation of regular maintenance procedures (see next section) or activities which made the loss or discard of items unlikely. A lamination of soot, grit, and other shoe-borne material of a thickness of up to 30 mm covered much of the parochial and eastern parochial burial areas, more particularly in the places which have been inferred as access points. The lamination is not a single deposit and we may infer that activity occurred many times over a long period. This would support a hypothesis that the bulk of the burial deposit was not introduced into the vaults at any one time.

In the post-excavation analysis interments were given a primary or secondary status. It is impossible to determine that a single deposit was primary on *priori* terms. A primary interment was defined as one which could not be demonstrated to have been culturally redeposited after its initial placement in a vault. A secondary interment was defined as one which exhibited demonstrable signs of having been culturally redeposited after its introduction into the crypt. The criteria used for ascribing secondary status to an interment were generally as follows:

(a) A *terminus post quem* existed which was later than the date of burial (date of death plus interval between death and burial where known). This apparently sound criterion relies on an assumption that coffins were not stored elsewhere in the crypt prior to their deposition in a vault, a possibility which cannot be discounted.

(b) The position of the skeleton within a coffin was disarticulated, indicating movement after a period of decay. Interments which may have been moved but would not exhibit identifiable disarticulation include those in which movement would not have been sufficiently violent for disarticulation to occur; those which were packed in sawdust, which would have prevented movement; and those for which decay processes were not sufficient to destroy tissue holding bones together. In this context articles by Brothwell (1987) and Boddington (1987) are especially interesting.

(c) The state of the coffin in comparison to those around it was so considerably worse that its condition was judged to be due to cultural redeposition. Only rarely was this last criterion used in isolation from other circumstantial evidence, since non-cultural destruction of coffin wood in the occasionally volatile environment of the crypt may be a significant factor. The present authors have already raised the possibility of gravity altering the culturally determined position of coffins (Adams & Reeve 1987, 252). The same process may account for the destruction of coffin wood without the need for human agency.

It is essential to account for these criteria because of the bias introduced into the analysis by their use. Many more interments than the number given secondary status in this analysis are likely to have been culturally redeposited at some stage. It may be valid, however, to use the figures produced here as an estimate of the minimum number of redeposited interments.

The effects of primary deposition on an interment and the mortuary environment with which it comes into contact are the most difficult events to reconstruct in the life history of a deposit. We have little evidence of the precise ways in which coffins were introduced into the vaults (see chapter 4). We may infer that certain deposits, such as fallen plaster and piles of damaged coffin wood, could have been generated during the period of primary deposition by, for example, a sexton and his helpers inadvertently banging coffins against walls. Anyone who has been through the trauma of watching removal operatives at work will vouch for the ethnographic analogy. Such primary and 'in transit' refuse was recovered from many parts of the crypt. We have almost no evidence for pre-depositional activities within the crypt. The Hurlin letter (see chapter 4.4; Appendix B) mentions that part of the burial service was held in the crypt, and that the relatives would have been present. The state of the vaults and the jumble of interments suggests that mourners would not have been encouraged to view the final resting place of the deceased. If indeed they were present at some subterranean stage of the burial service it seems likely that either a screen was placed between them and the open vault, or that some sort of antechamber was being used. Corroborative evidence for this important inference comes from a recent examination by the authors of the burial vault at St John's, Wapping. Here the division of the entire crypt into private chambers, many with viewing panels and almost all bearing inscribed stones, strongly suggests that mourners were allowed to perambulate around the crypt and look at the burials of their dead friends and relatives. Any other activities, such as the attachment of end plates, or the illicit removal of items of value may have taken place within the church. Evidence suggests (eg Curl 1972, 54) that the latter practice was rife, especially during the 19th century, and the presence of end or side plates on many coffins implies that these were being attached at or near the place of deposition. Certainly, the well-documented use of 'secure' coffins, of which there may be several examples at Christ Church, and the highly publicized practices of the 'sack-em-up gentlemen' (see chapter 4.2) indicates a potential recycling process, for which there does not seem to be any direct evidence at Christ Church.

Multiple burials in coffins, as well as interments excavated without receptacles, may offer some support to the thesis that coffins were being reused in some circumstances, but the depositional evidence would neither refute nor corroborate this. It is possible that primary interments were made which fell outside Christian liturgical practice during the period. Three interments which took place after 1809 appear to have been literally dumped within an abandonment phase deposit over the half-built blocking wall which sealed the lower portico vault; they were not contained in receptacles, but their positions indicate that the bodies were still articulated at the time of deposition.

An interment would have begun to interact with its environment immediately upon deposition. The extent of interaction would depend upon several factors. Some coffins were placed upon laths of wood, probably to facilitate removal of slings, but with the effect that the coffin would not immediately come into contact with the floor of a vault, inhibiting the effects of non-cultural processes such as damp (see section 5 below). Interments not contained in receptacles would have interacted at a greater rate with the environment, and been subject to different post-depositional processes, as described in section 5 and by Mant (1987).

Secondary deposition and maintenance

It is possible, as noted above, that all the interments excavated at Christ Church were secondary depositions according to the criteria listed in the section on primary deposition. That this is unlikely depends ultimately on an assessment of the stratification of the deposits, as outlined in chapter 3. An underlying assumption in the analysis of activities within the vaults at Christ Church is that settlement activities did not take place there. This apparent truism is important because it defines the range of processes which might be expected to leave traits behind for archaeological recovery. Nevertheless it should be possible to demonstrate the nature of the activities conducted in the crypt by reconstructing maintenance and discard processes operating there.

The traits and deposits produced by maintenance and discard activities should be easily differentiated from those produced by primary deposition. The rearrangement of coffins within vaults must have produced similar types of refuse to the primary activity. However, during periods of rearrangement maintenance processes were introduced, on an apparently random basis, producing deposits of secondary refuse. Piles of coffin wood fragments were located in many places. It is surprising in this context that no secondary refuse deposits of coffin furniture were located in the vaults. It is impossible to account for all of the coffin furniture which is assumed to have been present after primary deposition, and the authors infer that an additional process must have occurred which removed such metalwork from the area of burial. Such processes might include the recycling of metalwork, or its secondary disposal elsewhere.

During the autumn of 1987 the authors observed formation processes operating in the Terrace catacombs of Highgate cemetery in North London. The passages which contain burial recesses dating from the first half of the 19th century have been accessible to various parties during the last couple of decades. Here cultural processes include the desecration of coffins, the removal of some material, and the practice of a range of non-mortuary activities. The refuse patterns produced are in stark contrast to those resulting from the secondary depositional phase at Christ Church: vandalizing of memorials and coffins has produced distinctive primary refuse deposits, while maintenance activities occurring after such desecrations have resulted in neat piles of secondary refuse, especially of displaced coffin furniture, which are so conspicuously absent at Christ Church.

The authors were also permitted access, by the kindness of the Earl de la Warr, to the ancestral vault of the Sackville family at Withyam, in Kent. Here interments dating from the 16th century are contained in a purpose-built vault dating from the 18th century or later. Coffins have been redeposited periodically to create room for new interments, but maintenance activities here have been regular and thorough, with the result that very little primary or secondary refuse survives. Christ Church can conveniently be seen as something of a median between these two extremes. It displays both thorough maintenance activity, especially in a frequently used area such as the Lemaistre\Pontardant vault, and the apparently careless destruction and disposal of memorials, such as the Peck memorial which was broken and discarded within the dump of the Peck vault.

There may be two examples of the secondary discard of human remains not contained in coffins. In the lower north-west tunnel a group of nine skulls was recovered which had apparently originally belonged to complete interments in the crypt. In the lower central west tunnel a wooden box was excavated containing the disarticulated remains of parts of several individuals (0051). These two instances support the inference that occasionally during maintenance involving the re-placement of coffins some human remains would become separated from the original receptacle, becoming primary refuse which was then removed to a place of secondary refuse. It should be pointed out there there is no evidence of charnel activity at Christ Church.

An additional maintenance process, known from documentary evidence (Turner 1838) but only indirectly supported by the Christ Church data, concerns the activities of the so-called 'searchers'. These seem to have been individuals contracted either by an undertaker or directly by a family for the purpose of locating members of a family within a burial vault, so that new interments could be placed as closely as possible to others of the same familial grouping. It may be that the fixing of end and side plates to coffins whose breastplates would become invisible in a stack was conducted to facilitate this practice. At any rate, if such a practice was common it would have strong implications for social grouping at the site.

Abandonment

There is some evidence that burial areas at Christ Church were being abandoned at fairly regular intervals after the 1780s, and until 1866 or 1867. Schiffer defines abandonment as 'the process whereby a place . . . is transformed to archaeological context' (1987, 103). Archaeological context in this sense means the state of culturally deposited items in which they are no longer subject to human transformations, and are being operated on only by natural processes, or post-depositional processes. Systemic context is the state in which cultural processes operate on artefacts and deposits. An artefact can pass between the systemic and the archaeological contexts many times during its life history.

The burial area at Christ Church was certainly isolated from its systemic context after 1867, but parts of the crypt were evidently in a solely archaeological context long before then, perhaps as much as 70 years. The Lemaistre\Pontardant vault seems to have been abandoned shortly after 1795, by virtue of the space having been filled from floor to ceiling. After the closure of the second parochial vault around 1845 only the northern parochial vault and the south chasm seem to have been in use, and these were abandoned between 1852 and 1867. Precise dates of abandonment are difficult to ascertain, and it seems likely that in several cases abandonment dated by a *terminus post quem* from a coffin plate or inscription could be misleading.

An abandonment phase produces distinctive traits in the archaeological record, even if dating of its processes is equivocal. Catastrophic or unplanned abandonment is generally easily identified by the interruption of activities, large single deposits of materials, and especially by large amounts of *de facto* refuse. *De facto* is a term used to define the deposition of reusable items as refuse. The creation of *de facto* deposits implies the rejection of an item of cultural use, and of the area where the deposit occurs. Thus the deposit of, for example, half-eaten loaves of bread in the remains of a tavern in Pompeii strongly implies a rapid abandonment and rapid transformation of the area from the systemic to the archaeological context. If Christ Church had been abandoned in a hurry, we might expect to find reusable items belonging to the sexton and his or her helpers deposited in the crypt, such as unused candles, lifting straps, rollers, and even piles of cement and sand. However, we do not.

At Christ Church, then, judging by a singular lack of tools and other equipment deposited as *de facto* refuse, the abandonment was slow and planned. We may assume from knowledge of social values during the period of interment at Christ Church that the interments were not meant to be recovered at any stage; indeed, the deed for the sale of the Jarvis vault would support this, since the period of ownership of the space is defined as 'in perpetuity'.

There may, nevertheless, be a case for treating interments as if they were *de facto* artefacts, as discussed in chapter 4.5 and 4.6. The interments, at the point where a vault leaves the systemic context, might be seen to become *de facto* refuse, though this depends heavily on whether the interments and their receptacles are seen as reusable items. That the deposits have in the late 20th century been used by the scientific community for the purpose of the extension of knowledge suggests strongly that even in a social context like that of Victorian London, in which the dead were laid to rest 'in perpetuity', such cultural items are ultimately reusable. Certain areas of the crypt contain little if any *de facto* refuse. A small area beneath the steps leading down to the lower portico vault was filled with secondary refuse, the result of an apparently major clearance of broken coffin wood. The area under the portico steps themselves contained no refuse of any kind, and at present the crypt, emptied of interments and awaiting building work, is a repository for many kinds of materials, including archaeological planning frames, which have transformed the context of the crypt firmly from the archaeological to the systemic.

The abandonment of the burial areas at Christ Church is characterized by the sealing deposits which are present in most areas. Sealing was accomplished using several materials: building and demolition waste, as secondary refuse; sand and charcoal on top as a deposit of primary refuse with a probable sanitary function; and the bricks and mortar used in blocking entrances to the vaults in the western half of the church.

The introduction of the building and demolition waste can reasonably be dated to 1866 or 1867, when a Victorian restoration was being carried out on the main body of the church. The deposit includes plaster mouldings known to have decorated the ceiling above the nave, as well as rejected mouldings from the new interior. This deposit, which is homogeneous throughout most of the vaults, includes a wide variety of other materials. These include shoes, though no pairs of shoes, probably from a local cobbler's shop, leather offcuts, some sherds of ceramics, and many other items. The inventory closely resembles deposits of secondary refuse frequently noted by the authors in the vicinity of the church during the period of the excavations, and could be described in general terms as 'street rubbish'. Given the demonstrable continuity of economic activity in the vicinity of Christ Church during the last 250 years, it is hardly surprising that contemporary rubbish deposits bear such a striking resemblance to those used within the vaults. Of special interest to archaeologists is a single sherd of relatively unabraded samian ware recovered with fragments of 19th century domestic table wares in a 'dump' in the lower portico vault. The excavators infer that this is residual, but its presence in the crypt is open to interpretation. The introduction of the major part of the dump, which consisted of a matrix of earth, rubble, sand and mortar, totalling some 250 tonnes, may be supposed to have fulfilled two functions simultaneously in clearing the church of demolition refuse during restoration and filling the crypt according to the Queen's Council ruling of 1867 (see chapter 2.4). Thin layers of sand and charcoal were then deposited on top of the dump, before the vaults were finally bricked up.

Excavation

Henderson (1987) stresses above all that the major factors in determining the state in which human skeletal material reaches the human biologist are the limitations caused by the physical circumstances of recovery, and similar claims are made by Schiffer (1987) for the whole range of materials recovered by archaeologists. Christ Church conforms to this rule, for all-too-familiar reasons: rapid pace of recovery enforced by the pressure of time; unpleasant working conditions (see chapter 2.3); necessity for excavators to trample upon deposits prior to recording; poor lighting; fragility of material — the list is apparently endless, and does little to bolster the self-confidence of the archaeologist in the face of criticism from human biologists. It is worth, nevertheless, attempting to outline some of the major effects of the impact of archaeologists on the material recovered from Christ Church. The extent to which bias may have been introduced should then be clearer to the reader. It has been frequently noted that archaeologists in the field are wont to miss certain skeletal elements either because they are not looking for them — the smaller bones of the ear, for example — or because they are careless. Waldron (1987) demonstrated the relative survival rates of the different bones of the human body under varying conditions, and archaeologists can confirm from experience that bones such as the patellae are grossly under-represented in recovered skeletal samples. Unfortunately it is difficult even with the benefit of hindsight to determine whether differential recovery at Christ Church is due to differential decay or indifferent technique. A combination of the two is the most plausible explanation.

At the other end of the scale, the excavators took great care to ensure that, for example, a breastplate should not be associated with an individual unless it was firmly attached to a coffin and showed no signs of having been reattached. There were several instances in which two end plates were attached to a coffin, each bearing a different inscription. In such cases, unless one of these could be shown to have been a careless addition, neither plate would be ascribed to the individual. It is clearly evident that the introduction of an archaeological bias into this class of data would have had disastrous results for the human biological study. Consequently the dated interments represent a 'rump' of material which may be treated as unimpeachable as far as is possible.

The sampling strategy for the site has been outlined elsewhere in this volume (chapter 2.4), and its effect on the total recovered assemblage will be obvious. Metalwork was primarily selected either for biographical or taxonomic analysis, and items which were too corroded or fragmented to afford information on either of these criteria were not always retained. In similar fashion, during the later stages of the excavation emphasis was placed firmly on the recovery of skeletons with biographical information, and whilst every effort was made to ensure that equal attention was paid to all individuals interred in the crypt, the same cannot be said for the material culture associated with unnamed and undated interments.

An encouraging aspect of recovery at Christ Church is that the excavated area corresponds spatially for the most part with the area of interment. There were no archaeologically aesthetic sections cutting through interments, and the excavators could assume with confidence that there were no further interments lying just outside the excavated area to taunt them. With the exception of the Jervis vault and one or perhaps two other private interment areas, it is extremely unlikely that any interments are present in the crypt that have not been recovered.

5 Non-cultural formation processes at Christ Church

> 'Is there any other point to which you would wish to draw my attention?'
> 'To the curious incident of the dog in the night-time.'
> 'The dog did nothing in the night-time.'
> 'That was the curious incident,' remarked Sherlock Holmes.
>
> (from *Silver Blaze*, by Sir Arthur Conan Doyle)

As Holmes' remark implies, the absence of non-cultural disturbance at an archaeological site would be far more notable than, say, the complete destruction of entities by natural forces. However, as Boddington *et al* (1987, preface) stress, this state of affairs should not cause undue concern. Indeed, 'the analysis of the decay process itself can be harnessed for the purposes of reconstruction' (*ibid*).

Non-cultural, or natural, transformations are those processes which operate on archaeological material both before and after it has left the systemic context, but are especially important in the post-depositional, that is the archaeological, context. Mechanical, chemical, and biological processes which affect archaeological material are generally much better understood than they were even twenty years ago, and their effects are increasingly well documented (Boddington *et al* 1987; Schiffer 1987).

Materials subject to natural transformations will form the basis of the following discussion. As mentioned in section 2 above, the range of processes operating in the crypt environment can be subsumed under a few general headings. Since each of these processes affects different materials in particular ways, it seems appropriate to consider the effects of processes which have been dealt with comprehensively elsewhere (Wood & Johnson 1978; Schiffer 1987), rather than the physics of the processes themselves.

Human remains

The transformation of a human being to its eventual place in the lithosphere begins at the moment of death, and for the Christ Church individuals continues at present within the systemic context of the British Museum (Natural History). Cultural processes affecting the nature of transformation, such as the choice of receptacle and burial mode, the length of the post-mortem interval before burial, and the possible use of embalming techniques are dealt with in chapter 4. Their subsequent influence on the type and rate of non-cultural formation processes cannot be overstressed. Papers by Henderson (1987) and Mant (1987) illustrate this point powerfully.

Even for taphonomists the range of processes which operate upon the human body after the moment of death are bewilderingly complex. For the excavator to whom such expertise is unavailable the problem of recording post-mortem changes, let alone interpreting them, is one which has frequently resulted in ignoring the problem altogether (Garland 1987; 1988). Archaeological literature has until recently avoided the task of confronting such ignorance in the field, despite the fact that medical and palaeontological studies, though by no means profuse, have long been available.

The present authors have defined the parameters of their own ignorance rather too late in the day to affect recovery and field observation, but offer, in the hope of stimulating some practical changes in archaeologists' attitudes to dead bodies, a brief discussion of some post-excavational approaches.

A major problem appears to be the lack of a responsibility for retrieving the following data: condition of bone; type of preservation of tissue; formation of adipocere. Human biologists prefer to be given clean, unarticulated bones for their pathology and histology; and archaeologists, poorly equipped as they often, though not always, are to deal with organic remains, would rather that they did not have to deal with the problem. Somewhere in between a vast range of data is lost. The lament of the taphonomist is only now being heard (Boddington *et al* 1987).

The processes which may be expected to occur in a tomb or catacomb environment frequently result in many contrasting types of residue, and this has been recognized for more than 100 years. Lewis (1851) was contracted by the General Board of Health to consider the hazard posed by the interment of human remains in tombs within lead shells. This seems to have been prompted not only by the scare over the cholera epidemic in London, but also by a leader article in *The Times* of September 1849, which claimed that the unnatural obstruction of bodily decay processes caused by interment in lead was likely to incur a build-up of noxious gases within, which, upon explosion, would prove instantly fatal not only to those in the immediate vicinity, but also to the general populace.

Lewis, in rejection of this thesis, claimed to have visited all the major churches in London in order to observe for himself the conditions prevailing in tombs and crypts where interments had been deposited. This in itself is fascinating, for it raises the possibility that Lewis was observing bodily decay processes in the Christ Church crypt, as well as many others, 130 or so years before the intrusion of the excavators — sadly, we do not have any direct archaeological evidence for his presence. Lewis agreed that in principle, on both spiritual and health grounds, the practice of interments in crypts should cease, echoing the earlier study of Chadwick (1843). He observed a wide variety of processes affecting both receptacle and corpse, and was of the strong opinion that no coffin could be said to be completely sealed from the air. This opinion has been confirmed by Evans (1963), and by the excavators at Christ Church who have noted the presence of live insect larvae within apparently sealed coffins. He noted also the contrast, as did Evans, between bodies which retained a large amount of moisture, and those which were desiccated and appeared mummy-like. Evans' major point was to demonstrate that while the vapour escaping from lead-shelled coffins was extremely unpleasant, comparing it to 'very putrid, moist cheese', it was hardly fatal. Though he spoke to many sextons and their workmen, all of whom assured him that lead coffins would spontaneously explode, emitting a deadly gas which would burn with a blue flame, he noted that none had ever actually witnessed this phenomenon themselves. The excavators heard many similar stories from workmen claiming to have exhumed coffins from the Christ Church crypt during the 1960s, of which none could be corroborated.

The most frequently recorded post-mortem effects from Christ Church were diagenetic transformations of bone, desiccation of tissue and skin, complete skeletonization, and the formation of adipocere and brushite.

One of the most striking skeletal changes observed in the Christ Church interments was the high proportion in which the bone appeared to be composed of a moist but crystalline structure, had no solid core, and disintegrated upon retrieval to a dirty white-coloured 'mulch', probably brushite crystals. In almost all cases these were interments in wooden coffins without lead shells, and they were located for the most part in the lower portico vault, the south chasm, and the Peck vault. In all these areas coffin decay was extreme, the moisture content of the dump matrix was high, and the deposit of a considerable depth. It may be that the high moisture content contributed to an increase in hydrolysis, and it may be significant that the dumping material seems to have had a high lime content from plaster and cement. However, as Henderson (1987) notes, a complete understanding of the complexity of processes which may lead to this type of preservation\destruction is a distant goal. Unfortunately, figures relating to the relative survival and state of burials at Christ Church are not available. It appears that these details were not recorded by palaeopathologists working on the material. This issue has been raised in an article by the present authors (Adams & Reeve 1989f).

Generally the bone from Christ Church was well preserved in terms of its potential for pathological work. However, as Garland (1987, 109) notes, the terms 'well preserved' 'badly preserved', and 'weathered' are particularly unhelpful to taphonomists. It is considerably easier to identify agents of destruction than agents of preservation in a burial context such as that at Christ Church. Garland (*ibid*) stresses the contribution that serious study of the histology of archaeological human bone will be able to make to palaeopathology and forensic sciences in the future. This contribution includes tracing natural agents of transformation of human bone and other tissue. For the present, however, histological analysis of the Christ Church sample, which should have been much more rigorously controlled during the excavation stage, does not seem likely to aid in the development of that contribution. For this reason alone the examination of a sample of individuals from a similar archaeological context would be invaluable.

In areas of the crypt where dump deposits were not present, interments in wooden coffins often produced desiccation or mummification of the corpse. In these cases, notably many together in the northern parochial area, skin was preserved as a dry, parchment-like covering of the skeleton, beneath which tendons survived holding the skeleton together. Muscle and fat tissue was not observed, though there was some preservation of keratin, especially in finger- and toenails. A pinkish stain was observed on the bone in several cases. Frequently separation of the skin from the bone was difficult. None of the desiccated interments showed any signs of the formation of adipocere. No evidence of embalming was observed.

Evans (1963) notes that the belief that adipocere — a chalky wax-like substance generated from fat deposits — only forms when a body is subject to near waterlogging conditions is false. He stresses that sufficient water is usually available from the corpse itself for adipocere to form. Working with interments in a very similar context to those at Christ Church, Evans demonstrated that there was no necessary link between adipocere formation and embalming techniques, that there was a preponderance of adipocere formation in females, and that the nature of the environment during the post-mortem interval before burial, especially temperature and humidity, were important factors in the formation or otherwise of the adipocere deposits.

At Christ Church interments showing signs of adipocere were exclusively from coffins with lead shells, and generally from those in which body liquor was present. In many cases, especially in the parochial and eastern parochial vaults, in lead-shelled coffins which lay at an angle out of the horizontal, different types of decay were visible from the same interment. That part of the corpse remaining soaked in body liquor would often show signs of very complete preservation, including keratin, skin, muscle, and fat tissue, whilst that part of the corpse above the level of body liquor would generally be skeletonized (eg 2295). As Evans points out, coffins, whether of wood or lead or both, are very unlikely to be completely airtight. Thus such variation in the preservation from a single corpse can for the most part be ascribed to the relative effects of aerobic and anaerobic conditions.

Christ Church exhibits extremes of preservation and decay, most of which can be accounted for using the factors cited by Mant (1987) as the most important affecting rates and types of decay of the human body. The frequency with which coffins containing sawdust packing yielded clean, dry, and solid bones demonstrates the effectiveness of that kind of material for accelerating decomposition through the generation and\or conservation of heat during the early post-mortem period.

It is evident that just as factors affecting corpses prior to interment are difficult to account for, the wide range of micro-environments within the crypt area provide many clues to the ultimate retrieved state of the large majority of the interments. Those areas not containing large deposits of dumping material, and filled with primarily wooden-shelled coffins, provided a dry, stable mortuary environment with few major fluctuations in temperature, and apparently ideal conditions for desiccation. Other areas, such as the Peck vault and the south chasm, seem to have had dump material introduced during certain parts of the burial phase. These vaults are also on the south side of the church where there is known to be some seepage from a stream into the vaults. Consequently an environment damp from the leaking of body liquor and water seepage has produced a mortuary environment in which fungal, insect, and soil chemical processes have been encouraged, except where the anaerobic conditions of lead coffins containing substantial levels of body liquor have prevailed.

Though the environment of each vault was the primary determinant in the retrieved state of the corpse, interments were occasionally noted which bore evidence of different factors influencing decay and preservation. It has been inferred that in general those interments had been moved from one area of the crypt to another after the stabilization of decay processes was largely complete. However, it cannot be discounted that extreme conditions prevailing during the immediate post-mortem period, and indeed death itself, may in these cases be a major determinant.

Skeletons which were dislocated were generally accounted for in two ways, first by cultural disturbance after an initial period of decay (this process has been discussed above in the section on cultural formation processes), and second by excavation and retrieval techniques involving the movement of coffins prior to examination. Other cases can also be accounted for. Coffins which were initially buried upright, or were moved upright, usually contained a jumble of bones at the lower end of the coffin. Horizontal coffins displaying such a jumble were inferred to have been moved from the vertical to the horizontal position. One burial, 2607, exhibited bone disarray due to faunal turbation. Athough contained in an apparently sealed coffin lying under another coffin, examination revealed displacement of several of the left ribs. The cavity produced by this displacement contained some string fibres, a number of feathers, and the shredded remains of parts of more than one national daily newspaper, apparently from the second quarter of the 20th century. It was inferred that this deposit represents the remains of a rodent nest; rodent bones were present in several areas of the site. Careful reconstruction of the paper fragments should eventually yield a quite specific *terminus post quem* for this activity. Needless to say, it is a circumstance which may make archaeologists pause for thought before relying too heavily on the truism of Worsaae's law.

Wooden coffins

A review of processes of wood decay and preservation has been provided by Schiffer (1987). Major factors determining wood transformations are species of wood and the part of the tree, temperature and humidity of the environment, moisture, fungal and insect populations present in the environment, and working of wood. At Christ Church, coffin wood species were identified from gross observations. The representivity of species types is listed above in Table 5.2. No evidence of seasoning was observed, though it may be evident from analysis of the wood samples retained.

Wood was predominantly in three states. Wood remaining in a dry, stable environment was subject to little decay and was often still quite strong, frequently coinciding with corpses showing desiccation. Wood soaked continually in body liquor was extremely well preserved, and the liquor seems to have had some seasoning effect, for such wood was still very strong and virtually unbreakable. Most of the coffin wood had been subjected to attack by boring insects. Lewis (1851) notes the presence of the elm weevil (*rhyncolus lignarius*) in coffins in vaults contemporary with those at Christ Church. Where coffins had been attacked by boring insects the structure of the wood was usually severely destroyed and would crumble to dust upon being touched. Some wooden coffins in environments between these extremes exhibited predictable fungal and insect attack, which may in turn have affected the remains contained within the receptacle.

Calculations of load stresses for different types of wood and for lead have yet to be done for the Christ Church material. In general coffins were stacked up to seven high where there was sufficient vertical space. This did not seem to vary between single-, double-, and triple-shelled coffins. The effect of six coffins weighing up to a quarter of a tonne in total on a seventh at the bottom can be imagined. Wood, especially when moist, is transformed into little more than a pulp; lead shells are flattened to a thickness of as little as 100 mm, and consequently little of the corpse survives. In some cases, especially in the upper south tunnel, where there was very little dump deposit, lead-shelled coffins appeared to have been highly resilient to the stress imposed by stacking. Here up to six and even seven coffins stacked on each other exhibited only minimal crushing, and some of them remained effectively sealed until they were examined. It might be inferred that it was the added pressure of dumped rubble being introduced that contributed to the crushing that was widespread in the parochial areas.

Other wooden artefacts from the crypt seem to have fared poorly in the environment. Door frames and doors such as that separating the upper south tunnel from the south chasm crumbled to dust almost completely upon examination; lintels over doorways, of which three were examined, survived differentially. One had suffered extensive invasion by boring insects while the two others, still bonded into brickwork, had suffered only minor damage. No live boring insects were located during the excavations, and it may be the case either that excavation altered the environment sufficiently to discourage a population which had evidently thrived, or the conditions had been unfavourable for some time.

Lead

Besides the manufacture of coffin shells, lead was also used by the funeral industry during the Christ Church period to make coffin plates. Little is known about the specifics of lead coffin manufacture, although from the Christ Church data it has been possible to suggest a tentative taxonomy for their construction. The construction affects the length of time that a lead coffin will remain intact: a shoebox-type lid is, for example, much less secure than a coffin with flush-soldering along all the joints. Lead principally in the mortuary environment is destroyed by corrosion, forming carbonates which appear as crusts of 'white' lead. Such corrosion may destroy a coffin almost entirely, or render it brittle and weak. The process seems to act initially where the lead has been punctured by coffin plate nails, or along weak joints, and may be contributed to by decay products from the interment as well as moisture containing acids and other chemicals from the surrounding environment. Schiffer (1987) notes that lead is a corrosion-resistant metal except in circumstances where it comes into contact with organic acids.

8 Conclusion

The main justification for the excavation of the crypt at Christ Church, Spitalfields, has now been realized in the form of providing provenanced human material for an anthropological study. Apart from this anthropological study, further detailed analyses are being carried out on the collection, which has proved invaluable by virtue of the high percentage of aged and sexed individuals included in the group. In the area of historical research great steps have been made in the understanding of 18th and 19th century funerary practice, particularly that of the sexton's role and of the undertaking industry. Furthermore, the excavation has provided a collection of provenanced and closely dated funereal artefacts and clothing which offer a tight taxonomic group for any subsequent developments in the field. The survival of the textile artefacts and garments has been especially exciting for costume historians, besides offering a particular insight into the process of disposing of the dead in the period.

Many essential methodological lessons have been learnt and academic issues raised in the process of excavation and of writing this report, which further combine to emphasize the unique contribution of this site to archaeological understanding. The methodological lessons can be seen in two major areas, health and safety, and logistics. In chapter 2 a detailed catalogue of the health problems experienced and health restrictions imposed is given, which points up the contribution made by good physical health and good psychological morale to a successful project. Here, it is worth noting that archaeologists, like anyone else, thrive on natural sunlight and fresh air, and that the lack of them can seriously impede the performance of tasks; a factor that had not been taken into account. Concomitant with the negative aspects of the physical location of the excavation, it has been realized in retrospect that more time should have been allowed for the possible effects of lead contamination and the possibility of Health and Safety Executive and health authority intervention. Basic human welfare needs to be taken into account at the planning stages of a project so that overall timing allows for both sickness and health.

Another lesson which can be learnt from a retrospective assessment of the logistical problems faced in a crypt clearance is that excavation time cannot be condensed when all material — eg building material (*c* 245 tonnes), loam and charcoal deposits (*c* 5 tonnes), as well as coffins (*c* 900) sometimes weighing a quarter of a tonne — have to be removed by hand. What was originally estimated to be a six-month excavation expanded, with breaks of up to two months, into an excavation of eighteen months' duration. Some form of mechanization in the form of lifting devices or conveyor belts for spoil removal might have speeded up this process, although in most areas of the crypt restrictions of access would have prohibited their introduction.

The academic issues which have been raised by this project range from those thrown up by the analyses of the collected human material, through observations on the relationship between historical records of a period and their archaeological counterparts, to the more conceptual challenges of the nature of an archaeological context.

Insight has been gained into the working methods of four or five sextons who worked in the crypt between 1729 and 1867. The contrast between the stacking methods used in the parochial vault (PV) and the eastern parochial vault (EP) illustrate the differing individual impact of one worker over another very clearly. The comparison between the above-ground religious arrangements, as illustrated in the accounts of contemporary funerals (Appendix B), and the eventual disarray of the crypt after two major reshuffles in 1859 and 1867, further emphasize the care which must be taken when interpreting primary historical evidence. This particular issue brings up yet another taxing archaeological problem of interpretation: who buries the dead? The specialists who were in charge of procedures in the 18th and 19th centuries, the money-making undertakers, upholsterers, and coffin plate manufacturers, and the sometimes fastidious, sometimes slapdash sextons and their assistants, seem to have had the last word, although their concerns may have been far from those of the bereaved. In burial contexts without the benefit of historical comparison, the possibility that specialists, not the bereaved, have always buried the dead is highlighted by these findings.

It is necessary that the nature of an archaeological context be reassessed not so much as the remains of one activity in the past but as the residue of a combination of pre- and post-depositional impacts in the form of physical and chemical disturbances. The closely-dated nature of the archaeological context on this site has helped to define specific post-depositional changes. The maintenance processes of tidying up and creating good access through piles of coffins have been identified. Major reorganizations within the crypt can be tabulated minutely with the aid of the dates on the coffins. These dates have been used to help assess whether a coffin was in its 'primary' position or whether it had been moved to a 'secondary' position, as in the case of interment 2860, discussed in chapter 3.3 and 3.9.

Issues such as the recognition and recording of the effects of non-cultural formation processes or natural and environmental disturbances like disturbance by animals (faunal turbation) or gravity (graviturbation) have had to be dealt with on this excavation. The results of the subsequent analyses of these effects is made more universally applicable in consideration of the high number of known variables for the excavation. The composite parts of the site are known: the empty volume of the crypt, the mass of the infill, the time it took to fill the crypt, how it was done, and when it was sealed up and finally opened again. These components allow a clinical look at post-depositional activities which occur on every archaeological site.

It is important that the information derived from the excavation of the vaults at Christ Church, Spitalfields, is examined from two perspectives. Firstly, it represents the only large group of provenanced and closely dated human material and funereal artefacts from this period and therefore acts as a control to any further work in these areas. Secondly, the site acts as an analogue of a dirt site, where the component parts of a layer have been exploded to the size of a coffin and the effect of human trample above it has been magnified to the effect of 250 tonnes of rubble bearing down on those coffins, causing collapse and transformation. These contributions to the archaeological corpus have justified the excavation of Christ Church, Spitalfields.

Appendix A

References to Christ Church, Spitalfields, in primary documents

1 References to Spitalfields from the minutes of the building committee, 1711

485. 9 Nov 1711
Inhabitants of Spitalfields delivered plan of one and a half acre site and stated population of Hamlet at 18,000 or 20,000. There is one French church there, fit to be made parochial.

486. 13 Nov 1711
Commissioners to be recommended to build a second new church in Spitalfields ... [Brown's site]

2 References to Spitalfields from the minutes of Commissioners' meetings, 1714–27

91. 8 July 1714
8. Hawksmoor to take care that after foundations of Spitalfields new church is laid, a drain be made to carry off water into main sewer.
9. Resolved that there be two tiers of galleries in new church at Spitalfileds. [Only one tier of galleries was actually constructed on either aisle]

94. 29 July 1714
7. 250 to be allowed for planking foundation for Spitalfields new church.

114. 23 Mar 1714\15
2. Secretary to write to minister and churchwardens of Greenwich that the commissioners are informed that there have been several burials within the new church, contrary to the Acts, and that in future they forbear burying there.
3, 4. Upon the surveyors' complaint, resolved that Slemaker and Goodchild, bricklayers employed in new churches at Limehouse, Spitalfields and Upper Wapping [St George in the East], be discharged from any further service under this commission.
5. Surveyors to submit to next meeting proposals for getting better bricks.

154. 20 June 1716
4. Goodchild attended to answer complaint of his using bad bricks at Spitalfields.

190. 17 Oct 1717
6. Groves to make a shed at Spitalfields church for the engines [firepumps?], the place where they now are being spoilt by the new works.

210. 5 June 1718
10. Goodchild not to proceed with Spitalfields church until he provide such bricks as James [John James, Surveyor] approves of.

253. 28 Apr. 1720: The surveyors' memorial [memo]
3. The new church at Spitalfields has all its walls up to their proper heights and the two side ails [*sic*] covered with lead but in the middle of the church, although the walls are up and the timbers of the roof prepared, yet the carpenter has not bin [*sic*] able to compleat the said roof for want of money, having a great debt, at this church and in other parishes, due to him, of above £2,000. ...
7. We humbly move that the ground brought [*sic*] for sites of the churches of Limehouse, Spitalfields, Wapping and Westminster may be walled in with brick, to prevent the harm continually done by the mob to the buildings and works.

373. 27 July 1724
4. The two rectors of Stepney attending were called in and heard concerning districts for the new parishes. Ordered that entire hamlet of Spitalfields be parish of the new church.

396. 15 Feb 1724\25
2. [Endowments]
Large parishes and small rents and many poor: Westminster, Spitalfields and Wapping,

by a rate yearly for each church	270
by the parliamentary fund	80

whereof the curate £80, the residue to the rector.

456. 16 Mar 1726\7D
3. Whereas Spitalfields offered to settle £200 p.a. for minister's maintenance it now appears that £150 p.a. is as much as can be expected because of decay of trade and fall of rents.

3 From the building records [Lambeth Palace Library MS 2698, 41]

25 March 1722–25 March 1723

[Account to Richard Goodchild, bricklayer, includes:]

8 Brickwork done in the Ten Inner Peers of the Vaulting...

... 15yds cub. of digging at pr yard 0-0-6

... Daywork done by him in digging and carrying out of the vaults of the church and levelling it about ye church yard 0.7.6

102 days of a labourer at 0-1-8 8-10-0

4 Orders in Queen's Council relating to burials at Christ Church

(1) At the Court at *Buckingham Palace*, the 20th day of *March*, 1859,

PRESENT

The QUEEN's Most Excellent Majesty in Council.

Whereas by an Act, passed in the session of Parliament held in the fifteenth and sixteenth years of Her Majesty's reign, intitled 'An Act to amend the laws concerning the burial of the dead in the metropolis'; it is enacted, that in case it appears to Her Majesty in Council, upon the representation of one of Her Majesty's Principal Secretaries of State, that, for the protection of the public health, burials in any part or parts of the metropolis, or in any burial-ground or places of burial in the metropolis, should be wholly discontinued, or should be discontinued subject to any exception or qualification, it shall be lawful for Her Majesty by and with the advice of Her Privy Council, to order that from and after a time mentioned in the Order, burials in such part or parts of the metropolis, or in such burial-grounds or places of burial, shall be discontinued wholly, or subject to any exceptions or qualifications mentioned in such Order, and so from time to time, as circumstances may require; provided that notice of such representation, and of the time when it shall please Her Majesty to order the same to be taken into consideration by the Privy Council, shall be published in the London Gazette, and shall be affixed on the doors of the churches or chapels of the parishes in which any burial-grounds or places of burial affected by such representation, shall be situate, or on some other conspicuous places within the part or parts of the metropolis affected by such representation, one calendar month at the least before such representation is so considered; provided always, that no such representation shall be made in relation to the burial-ground of any parish until ten days' previous notice of the intention to make such representation shall have been given to the Incumbent and the Vestry Clerk of such parish:

And whereas the Right Honorable Sir George Grey, Bart., one of Her Majesty's Principal Secretaries of State, after giving to the Incumbents and Vestry Clerks of the Parishes hereinafter mentioned ten days' previous notice of his intention to make such representations, has made two representations stating that, for the protection of the public health, burials should be discontinued in the hereinafter-mentioned places:

Now, therefore, Her Majesty, by and with the advice of Her Privy Council, is pleased to order, and it is hereby ordered, that burials be discontinued in the under-mentioned places, from and after the first of April next; viz:

[as well as nine other churches]

CHRIST CHURCH, SPITALFIELDS. — Beneath Christ Church, Spitalfields, and also in the vaults under the entrance steps and porch of the said church.

(2) At the Court at *Buckingham Palace*, the 18th day of *April*, 1859,

CHRIST CHURCH, SPITALFIELDS

That the ventilation gratings connected with the vaults beneath Christ Church, Spitalfields, and openings on the north and east side of the church, be closed with stone or brickwork.

(3) At the Court at *Buckingham Palace*, the 6th day of *June*, 1859,

Now, therefore, Her Majesty, by and with the advice of Her Privy Council, is pleased to order, and it is hereby ordered, that burials be discontinued, from and after the fifteenth day of this instant June, in the said *churchyard* of CHRIST CHURCH, SPITALFIELDS.

(4) At the Court at *Osbourne House, Isle of Wight*, the 2nd day of *February*, 1867,

Now, therefore, Her Majesty, by and with the advice of Her Privy Council, is pleased to order, and it is hereby ordered, that the churchwardens or such other persons as may have care of the undermentioned church do adopt, or cause to be adopted, the following measures, viz:

That the coffins now unenclosed beneath the parish church of Spitalfields, Middlesex, be embedded in soil mingled with charcoal, and enclosed either by concrete or by stonework or brickwork properly cemented.

Appendix B

Accounts of funerals at Christ Church

1 Account of the funeral of Mrs Sarah (Marchant) Hurlin, who died May 6, 1839, written by special request of NMH [?National Museum of History]

My grandmother had a special desire for what was called 'a grand funeral' which was quite expensive in those days, and it was understood that she laid aside a sum of money, perhaps $500.00 for this special purpose. I think also that she shrank from the idea of being *buried in the earth*.

At the parish of Spitalfields, about three quarters of a mile from her home in Bethnal Green, the whole space under the building was laid out as a burial vault, and it was here that she wished to be buried. As it was in a populous neighbourhood, it was necessary to guard against effluvia, and thus in order to be buried in this vault it was required that three coffins should be used, first one of oak, in which the body was deposited; this was placed in a metallic coffin, on which the lid was closely soldered, and this was placed in a common coffin, covered with black woollen cloth, and with a row of black nails with large heads placed close together around the sides and ends of the coffin, and around the edge and top of the lid. There were also handles and other ornaments on the sides and a plate containing name and age on the lid.

On the day of the funeral, an hour before it occurred, the two men called 'mutes' took their places outside the front door, one on each side of it. They were dressed in black, and had a black silk sash over one shoulder, and across the breast low down on the other side. Each had a broad black silk band on his hat with the ends of it, about 18 inches long, hanging down behind. In the hand farthest from the door, they each held a black staff about four and a half feet long, with a cross piece about a foot long on top of it, over which was laid a piece of black silk about eight feet long, which was bound close to the staff about two feet from the top, and its folds hung loose about two feet lower.

When the mourners were assembled, the undertaker dressed them, the men in long black cloth coats and black gloves, and a broad black crepe band around their hats, the ends of which hung down about 18 inches; and the women in a long black scarf with a hood attached to it which covered their heads, and also with black gloves. When all were ready the procession started as follows:

First, the undertaker in black with a black silk scarf and black silk hat band like the mutes.

Second, the mutes aside each other with their staves.

Third, a man in black with a black board about three by one and a half feet, on his head, on which board were placed six groups of four or five large black ostrich feathers.

Fourth, the hearse covered with black cloth and drawn by two black horses, each having on his head a bunch of large black ostrich feathers; and eight groups of large black ostrich feathers on top of the hearse in which of course the coffin had been placed. The driver was dressed in a black cloak with silk hat band hanging down behind, like the undertaker and the mutes. An attendant in black with sash and hat band like the mutes, walked outside each horse, and each of them carried a small staff about 18 inches long which he grasped in the middle of it.

Fifth, two mourning coaches to accommodate six persons each, each horse having a plume of feathers on his head, the drivers being dressed as the driver of the hearse, and there being two attendants for each coach as in the case of the hearse.

After the prescribed portion of the Burial Service had been read in the church, the coffin was carried into the vault, and deposited in the place assigned it, and the remainder of the service was read. I think the attendants before spoken of acted as bearers.

As the oldest son of the oldest son, and also the oldest grandchild, I attended the funeral as the representative of the grandchildren.

Wm. Hurlin
Antrim, N.H. [New Hampshire]
September 30th, 1908

2 Description of the funeral of Colonel Paul Le Mesurier from the obituary in the *Gentlemen's Magazine*, 1806, vol 76 pt 1, pp 84–6

. . . On the Friday he was with difficulty persuaded to call in his carriage upon Dr Babington, who immediately said that the case was beyond the aid of medicine. On the Saturday, Sunday and Monday he kept at home, but not in his bed, and even partook of a light dinner, on the latter day, at three o'clock, but expired in his chair at about ten. As soon as the event was known, every desire was most ardently and honourably evinced to pay the last respectful attention to his remains by the several public bodies with whom he was connected.; the Court of India Directors, the three India-house Volunteer Regiments, the Hon. Artillery Company, the City-Marshalls and Household Officers, and the Hackney Volunteers, all severally offered to attend his funeral; but, privacy being the wish of the family, the same was handsomely declined. This negative could not, however, prevent the warmth of tributary feeling shewn by his own regiment, the Hon. Artillery Company, who, to avoid intruding upon the sacredness of the domiciliary wish, repaired from the Artillery-house, near 700 strong, to Clapton-field (an open space, quite contiguous to Upper Homerton, the residence of the deceased), and, being drawn up in marching order, led the van of the solemn cavalcade (leaving their field-pieces upon the ground to fire minute-guns during two hours). Sir John-William Anderson, and Sir William Curtis, barts. the president and treasurer of the Company, attended by the surgeon, secretary, and two other civil officers of the same, headed the regiment in two mourning-coaches drawn by four horses each, richly plumed, and surrounded by a party of Yagers. The regiment followed, in full military order of march, the band playing solemn dirges on crape-hung instruments. The charger of the deceased, bearing his arms and accoutrements reversed, was led immediately before the hearse, which was followed by four mourning-coaches, each conveying four friends or relatives. The procession was almost over-pressed by the concourse of spectators on its way to Christ Church, Middlesex (Spitalfields), where it was received by the Parochial Volunteers, expressly drawn out. After it entered the church, under the solemn pealing of the organ, the service was most excellently performed by the officiating clergyman; and, the body being deposited in a family vault under the church, six volleys of musketry were fired in the church-yard, as the conclusion of the last sad respectful ceremony to a man whose death will long remain both a public and private loss . . .

Appendix C

Inscriptions on memorials in the church

1 Inscriptions on memorials in the church dated between 1729 and 1867 which directly refer to burial within the crypt, in chronological order (see Fig 1.4)

Note *Those which were removed in 1981 during the restoration programme and are in temporary storage are marked* *

1 HERE LYETH INTERR'D THE BODY OF
ISAAC LEFEVRE SON OF MR ISAAC &
MRS JUDITH LEFEVRE OF THIS PARISH
WHO DIED JANUARY Ye 2d 1730
AGED 9 MONTHS & 5 DAYS.
HERE ALSO LYETH INTERR'D THE BODY OF ANNE
LEFEVRE DAUGHTER OF Ye ABOVE SAID WHO DEPARTED
THIS LIFE THE 10th DAY OF OCTOBER 1732 AGED 7
YEARS 11 MONTHS & 21 DAYS.
HERE ALSO LYETH SARAH LEFEVRE OF Ye ABOVE Sd
WHO DIED AUG Ye 15 1739 AGED 8 YEARS 11 MONTHS
AS LIKEWISE THE BODY OF MR ISAAC LEFEVRE
WHO DEPARTED THIS LIFE MAY 12th 1746
IN THE 61st YEAR OF HIS AGE
AND ALSO THE BODY OF MRS JUDITH LEFEVRE
WHO DEPARTED THIS LIFE THE 15th OF APRIL 1780
IN THE 85th YEAR OF HER AGE.
SACRED ALSO TO THE MEMORY OF JOHN LEFEVRE OF OLD FORD
AND OF HECKFIELD PLACE IN THE COUNTY OF SOUTHAMPTON
ELDEST SON OF THE SAID ISAAC LEFEVRE AND JUDITH HIS WIFE
WHO DIED JANUARY 16 1790 AGED 69 AND WAS BURIED AT WEST HAM
AND OF HIS WIDOW HELEN LEFEVRE DAUGHTER AND CO-HEIRESS OF
LISTER SELMAN
OF OLD FORD & CHALFONT St GILES, BUCKS WHO DIED APRIL
1816 AGED 81, & WAS BURIED AT
WEST HAM. ALSO OF THEIR ONLY CHILD & HEIRESS HELENA
SHAW-LEFEVRE
OF OLD FORD & HECKFIELD PLACE WHO DIED AT MONTI VILLIERS,
NEAR HAVRE FRANCE
AUG 17 1834 AGED 67 AND WAS BURIED HERE. ALSO OF HER HUSBAND
CHARLES SHAW-LEFEVRE M.A. F.R.S. LORD OF THE MANOR OF BURLEY
FOR 18 YEARS M.P. FOR READING WHO ASSUMED THE SURNAME AND
ARMS OF LEFEVRE
BY ROYAL LICENCE ON HIS MARRIAGE. HE DIED APRIL 1823 AGED 63
ALSO OF THEIR SONS CHARLES, VISCOUNT EVERSLEY C G B
SPEAKER OF THE HOUSE OF COMMONS; GOVERNER OF ISLE OF WIGHT
HIGH STEWARD OF WINCHESTER; YEOMANRY ADC TO THE QUEEN
HE DIED AT HECKFIELD PLACE DEC 28 1888 AGED 94
SIR JOHN GEORGE SHAW-LEFEVRE KCB SENIOR WRANGLER IN THE
UNIVERSITY OF
CAMBRIDGE; CLERK TO THE PARLIAMENTS SOMETIME
MP FOR PETERSFIELD

HE DIED AUG 20 1879 AGED 82
HENRY FRANCIS SHAW-LEFEVRE MERCHANT AND BANKER IN THE CITY OF LONDON HE DIED DEC 3 1880 AGED 78
ALSO OF HELEN DAUGHTER TO MAJOR GENERAL JOHN CASPARD LE MARCHANT
SISTER TO SIR DENIS LE MARCHANT [?] BARONET AND FIRST WIFE OF HENRY FRANCIS SHAW-LEFEVRE. SHE DIED FEB 19 1833 AGED 30 & WAS BURIED HERE.

2 IN A VAULT AT THE SOUTH WEST END OF
THIS CHURCH IS DEPOSITED THE BODY OF
EDWARD PECK ESQ, WHO DEPARTED
THIS LIFE THE 19th OF JUNE 1736 IN THE
75th YEAR OF HIS AGE.
HE WAS ONE OF THE COMMISIONERS FOR
BUILDING THE FIFTY NEW CHURCHES AND
IN THIS CORNER LAID THE FIRST STONE
OF THIS STATELY FABRICK ANNO 1715
IN THE SAME VAULT IS DEPOSITED THE
BODY OF ELIZABETH WIFE OF THE
SAID EDWd PECK WHO DEPARTED
THIS LIFE THE 25th OF JULY 1730 IN
THE 62 YEAR OF HER AGE.
LIKEWISE THE BODY OF MRS DEBORAH
LATE WIFE OF JOHN PECK ESQr WHO
DEPARTED THIS LIFE THE 26th OF
NOVEMBER 1739 IN THE 36th
YEAR OF HER AGE
ALSO THE BODY OF JOHN PECK ESQr
WHO DEPARTED THIS LIFE ON THE 14th
DAY OF MARCH 1748 IN THE 48th YEAR
OF HIS AGE
HE WAS THE LAST MALE ISSUE OF Ed PECK ESQr
WHOSE EXCELLENT VIRTUES HE IMITATED AND
WAS BOTH THE CHRISTIAN AND THE GENTLEMAN
TO WHOSE MEMORY THIS
MONUMENT IS ERECTED BY
THEIR SURVIVING SON JNo PECK
ANNO 1737

3 SACRED TO THE MEMORY
OF SIR ROBERT LADBROKE KNT AND ALDERMAN
PRESIDENT OF CHRISTS HOSPITAL
AND OF THE ANNIVERSARY OF THE CHARITY SCHOOLS IN
LONDON AND WESTMINSTER;
LORD MAYOR IN MDCC XLVII
WHICH OFFICE HE EXECUTED SO MUCH TO HIS CREDIT
AND TO THE SATISFACTION OF HIS FELLOW-CITIZENS
THAT THEY FIXED UPON HIM AT THE NEXT GENERAL ELECTION
FOR ONE OF THEIR REPRESENTATIVES IN PARLIAMENT:
WHICH IMPORTANT TRUST THEY CONTINUED HIM TILL HIS DEATH.
HE WAS A MAN OF A MOST MILD DISPOSITION;
DISINTERESTED IN HIS VIEWS; OPEN AND GENEROUS IN HIS CONDUCT:
AND SO INDEFATIGABLE IN PROMOTING THE TRUE INTERESTS
THAT HE VERY JUSTLY MERITED HIS TITLE OF
FATHER OF THE CITY.
AS HE LIVED GREATLY ESTEEMED BY ALL WHO KNEW HIM

HE DIED MUCH LAMENTED OCT XXXIst MDCCLXXII AGED LXI YEARS
AND LIES BURIED IN A VAULT UNDER THIS CHURCH
WITH HIS AMIABLE LADY ELIZABETH. DAUGHTER OF JOHN BROWN ESQ
AND GRANDDAUGHTER OF JOHN PECK ESQ LATE OF THIS PARISH
WHOSE PROPERTY SHE CHIEFLY INHERITED
AND DIED UNIVERSALLY REGRETTED OCT 1st MDCCLXVIII
AGED LIII YEARS
RICHARD LADBROKE LATE OF FRENCHES IN THE COUNTY OF SURREY ESQr
BY HIS WILL ORDERED THIS MONUMENT TO BE ERECTED.
AS A TRIBUTE OF RESPECT AND GRATITUDE
DUE TO SO WORTHY A MAN, SO GOOD AN UNCLE
AND SO FAITHFUL A GUARDIAN
MDCC XCIV

4* IN MEMORY OF
MR PHILIP CHABOT OF THIS PARISH WHO DIED OCT 6 1800 AE 57
LYDIA HIS WIFE WHO DIED MARCH 18 1801 AE 53
GEORGE HIS SON WHO DIED JANUARY 26 1796 AE 14
ELIZABETH HIS ELDEST DAUGHTER WHO DIED MAY 30 1813 AE 40
PHILIP HIS ELDEST SON WHO DIED NOVEMBER 2 1832 AE 57
CHARLOTTE HIS DAUGHTER WHO DIED SEPTr 9 1840 AE 66
THE ABOVE ARE INTERRED
IN THE FAMILY VAULT IN THE CHURCHYARD ADJOINING
ALSO OF
JAMES CHABOT SECOND SON WHO DIED JUNE 1850 AE 72
HARRIET NÉE BEEK HIS WIFE WHO DIED DECEMBER 31 1850 AE 70
WHO BOTH DIED IN THE CITY OF MEXICO
EMILY SEMPRONIA DAUGHTER OF JAMES AND HARRIET WHO DIED MAY 31 1828 AE 2
INTERRED IN HER GRANDFATHERS VAULT
ALSO OF PHILIP JAMES CHABOT ESQr M.A. F.R.A.S. ELDEST SON OF
THE LAST NAMED PHILIP CHABOT AND ELIZABETH HIS WIFE
NÉE PARNELL WHO DIED JANy 11 1868 AE 66 HE IS INTERRED IN
ABNEY PARK CEMETERY IN THE VAULT OF ROBERT WESTWOOD ESQre

5 SACRED TO THE MEMORY OF
CHARLES SMITH ESQUIRE OF STRATFORD
IN THE COUNTY OF ESSEX
HE DIED ON THE VIII DAY OF FEBRUARY MDCCLXXVII
IN THE LXIV YEAR OF HIS AGE
ALSO TO THAT OF JUDITH HIS WIFE
WHO DIED THE XVII DAY OF MARCH MDCCCVIII
AGED XCI YEARS
THEIR REMAINS TOGETHER WITH THOSE OF
SARAH THEIR ELDEST DAUGHTER
WHO DIED ON THE XVII OF NOVEMBER MDCCLXXVII
AGED XXII ARE DEPOSITED NEAR THIS SPOT
IN THE FAMILY VAULT OF ISAAC LEFEVRE ESQ
FORMERLY OF THIS PARISH AND FATHER OF
THE ABOVE NAMED JUDITH SMITH
THIS TABLET IS ERECTED
WITH THE GREATEST FILIAL REVERENCE
BY THEIR TWO SURVIVING CHILDREN
IN THE SAME VAULT ARE INTERRED THE REMAINS OF
JUDITH

YOUNGEST DAUGHTER OF THE ABOVE
CHARLES & JUDITH SMITH
WHO ENDED A LIFE OF
UNCEASING BENEVOLENCE AND PIETY
25th FEBRUARY 1832
AGED 77

6* IN THE SANDELL VAULT BENEATH THIS
SPOT ARE DEPOSITED THE REMAINS OF
MARY
WIFE OF SAMUEL SANDELL ESQr
OF OXFORD STREET
WHO DIED OCT 12th 1814
AGED 61 YEARS
ALSO
THE ABOVE NAMED SAMUEL SANDELL ESQr
WHO DIED JULY 24th 1836 AGED 61 YEARS

7* IN THE FAMILY VAULT OF THE LATE
W C HEADINGTON ESQ OF THIS PARISH
ARE INTERRED THE REMAINS OF
HIS GRAND DAUGHTER
MARY ANN HOLLOWAY
WHO DIED ON THE 27th DEC 1829
IN THE 17th YEAR OF HER AGE
BELOVED, ESTEEMED AND LAMENTED
THIS TABLET IS ERECTED TO HER MEMORY
BY HER AFFECTIONATE MOTHER
HARRIETT HOLLOWAY
RELICT OF J P HOLLOWAY ESQre
THE REMAINS OF HARRIETT HOLLOWAY
ARE INTERRED IN THE CHURCH OF
TODDENHAM St MARTIN, SUFFOLK
SHE DIED 27th 1857 AGED 80
'I KNOW THAT MY REDEEMER LIVETH'

8* SACRED
TO THE MEMORY OF
JOSEPH VAUX
WHO DEPARTED THIS LIFE MAY 13th 1835
AGED 71
ALSO
MARY VAUX WIFE OF THE ABOVE
WHO DEPARTED THIS LIFE DECEMBER 10th 1835
AGED 72

2 Inscriptions on memorials which were found in the crypt or found in dumps in the crypt, in chronological order.

*Those which have since been moved from their original provenance are marked **

9* THIS WALL WAS ERECTED AT THE
CHARGE OF THE PUBLIC IN FULL
SATISFACTION FOR 68 FEET OF
GROUND EAST OF THE SAME
WHICH WAS CONVEYED TO THE
HAMELET OF SPITTALFIELDS

FOR EVER
BY CHARLES WOOD
AND SIMON MICHELLSON
IN THE YEAR OF OUR LORD 1727

10* IN THIS VAULT LIES INTERRED
MARY FOLWELL
Obt 6th JUNE 1729 aet 65
ELIZABETH FOLWELL
Obt 3rd AUGUST 1741 AET 46
JOHN FOLWELL
Obt 24th AUGUST 1744 AET 76
OUs WHETHER IN ORDER OF
TIME OR OF RELATIONS

11* WITHIN THIS STONE IS
DEPOSITED THE BODY OF
Mrs ELIZABETH PECK
LATE WIFE OF
EDWARD PECK ESQ
OF THIS PARISH WHO
DEPARTED THIS LIFE THE
25th OF JULY 1730
IN THE 62nd YEAR
OF HER AGE

12 THIS VAULT NO. 10 WAS PURCHASED BY MR DANIEL MESMAN OF
THIS PARISH JULY THE 27th 1731
IN Ye SAID VAULT LYES Ye BODY OF Mr JOHN
MESMAN BROTHER TO YE ABOVE S'D Mr DANl.
MESMAN WHO DIED JULY Ye 23 1731 AGED 61 YEARS
ALSO Ye BODY OF MARTHA MESMAN DAUGHTER
OF DANl & MARTHA MESMAN WHO DIED SEPTbr
THE 13th 1730 AGED 6 MONTHS 28 DAYS.
ALSO Ye BODY OF Mr DANIEL MESMAN
WHO DIED FEB Ye 4th 1733 AGED 70 YEARS.
ALSO Ye BODY OF MARTHA MESMAN WHO
DIED AUGst Ye 21st 1733 AGED 4 MNTHS 10 DAYS
ALSO Ye BODY OF MARY MESMAN WHO DIED
JULY Ye 31st 1734 AGED 1 MONTH 21 DAYS
ALSO Ye BODY OF JANE MESMAN DAUGHTER
OF JOHN AND DINAH MESMAN WHO DIED
AUGT Ye 29th 1735 AGED 5 WEEKS
ALSO Ye BODY OF MARY MESMAN DAUGHTER
OF DANl. & MARTHA MESMAN WHO DIED
JULY Ye 9th 1737 AGED 8 MONTHS 3 DAYS.
ALSO Ye BODY OF Mr JOHN MESMAN
WHO DIED DECEMBER Ye 27th 1737 AGED 32 YEARS.
ALSO Ye BODY OF Mrs JANE MESMAN WIDw
OF Ye ABOVE S'D DANL. MESMAN WHO DIED
APRIL Ye 4th 1739 AGED 73 YEARS
ALSO Ye BODY OF SAMUEL MESMAN WHO
DIED MAY Ye 10th 1740 AGED 16 MTHS 7 DAYS.
ALSO Ye BODY OF SAMUEL MESMAN WHO
DIED OCTbr Ye 17th 1740 AGED I MONTH 13 DAYS
MARTHA MESMAN DIED THE 20th
MAY 1754 AGED 48

DANIEL MESMAN ESQ DIED
THE 12th MAY 1765 AGED 65
JOHN MESMAN DIED
SEPTbr 11th 1768 AGED 37
MARY MESMAN DIED
MARCH 3rd 1772 AGED 44
CHARLES MESMAN DIED
MARCH 14th 1775 AGED 34

13* ENTOMBED ARE THE REMAINS OF
EDWARD PECK ESQ
WHO DEPARTED THIS LIFE THE
19th OF JUNE 1736
AGED 75 YEARS
HE WAS ONE OF THE
COMMISSIONERS FOR BUILDING
THE FIFTY NEW CHURCHES AND
LAID THE FIRST STONE OF THIS
STATELY FABRICK IN THE
SOUTH EAST CORNER
ANNO DOMINI
MDCCXV

14 THIS VAULT WAS PURCHASED BY Mr EDw CARTER
[PLUMBER?] OF THIS [PARISH?] APRIL Ye 10th 1744 IN IT LIES
EDw CARTER WHO DIED FEB Ye 17 1738 AGED [?] YEARS & 6 MONTHS
JOHN CARTER DIED MARCH 26th 1739 AGED 8 YEARS & [?] MONTHS
THO CARTER DIED JUNE [?] 1743 AGED 4 YEARS NINE MONTHS
ALSO ANN CARTER MOTHER OF THE ABOVE Mr EDWARD CARTER
DIED FEB Ye 27th 1756 AGED 96 [?] YEARS
ALSO Mrs MARY CARTER LATE WIFE OF Ye ABOVE NAMED EDWARD CARTER
WHO DIED MAY Ye 22 1762 AGED 54 YEARS

15 THIS VAULT WAS PURCHASED BY LEONARD SNEE APRIL Ye 1st 1740
IN IT LYETH
MARY SNEE WHO DIED MARCH 29th 1740 AGED 19 YEARS 6 MONTHS
LEONARD SNEE WHO DIED FEBRUARY 14th 1766 AGED 73 YEARS
MAGDALENE SNEE DIED 21st JANUARY 1774 AGED 79 YEARS
ALSO PERCIVAL BARKER DIED JUNE 6 1786 AGED 58 YEARS
ALSO SUSANNAH BARKER DIED FEBr 6th 1805 AGED 74 YEARS
MARY BARKER SPINSTER DIED OCT 27th 1808 AGED 46 YEARS
ALSO ANN SNEE DIED FEBRUARY 20th 1819 AGED 60 YEARS
JANE BARKER DIED 16th OCTOBER 1820 AGED 55 YEARS
ELIZ BARKER DIED 28 MARCH 1822 AGED 67 YEARS
ELEANOR BARKER DIED 10th SEPTEMBER 1826 AGED 70 YEARS

16 THE VAULT OF Mr PETER
JAMES DOUXSAINT
CITIZEN AND CLOTHWORKER
OF LONDON MERCHANT HE
BOUGHT THE SAID VAULT
FOR HIS FAMILY JANUARY
THE 4th 1741/2

17* JOHN PECK ESQ

DIED MARCH 20th 1749
IN THE 49th YEAR
OF HIS AGE

18 THIS VAULT BELONGS TO
Mr JAMES LEWIS DESORMEAUX
AND THE LATE MR JOHN GILSON
PURCHASED OF THE PARISH
THE 9TH OF APRIL 1786
HENRY NELSON CHURCHWARDEN

19* THE FAMILY VAULT OF
Mr STEPHEN WILSON
AND
Mr WILLIAM WILSON
FEB 27th 1791

20 THE NEXT VAULT
UPON THE LEFT BELONGS TO
Mr DANIEL VERNEZOBRE
.... MERCHANT WHO
.... FOR HIS FAMILY
..... 1117.. ......
.... FOR THE SAME

21 THIS VAULT WAS ERECTED BY ORDER OF THE VESTRY 1813
REV WEST WHELDALE RECTOR
THOMAS LOW
GEORGE WALKER CHURCHWARDENS

22* THE VACANT GROUND EXCLUDING
61 FEET EAST 13 FEET WEST AND
30 FEET NORTH OF THIS STONE IS
PART OF THE GROUND CONVEYED
TO THE COMMISSIONERS APPOINTED
BY ACT OF PARLIAMENT
FOR BUILDING THE CHURCH
AND IS ACCORDINGLY PART OF THE
FREEHOLD BELONGING TO THE SAME
REVd WEST WHELDALE RECTOR
WILLIAM BARTHOLEMEW
WILLIAM DAVES CHURCHWARDENS
1821

23 HERE ALSO REST THE MORTAL REMAINS OF
JOHN SNEE WHO DIED 17th JULY 1827 AGED 73 YEARS
CATHERINE BARKER 13 APRIL 1833 AGED 75 YEARS
SARAH SNEE 12th MARCH 1835 AGED 78 YEARS
AMELIA BARKER 2nd JANUARY 1839 AGED 75 YEARS
JANE EMILY SNEE 8th NOV 1839 AGED 39 YEARS
ELIZABETH SNEE WIDOW OF THE ABOVE NAMED
JOHN SNEE DIED 7th MAY 1842 AGED 80 YEARS
SOPHIA BARKER DIED 21st AUGUST 1842 AGED 80 YEARS
LYDIA BARKER DIE 26th FENRUARY 1847 AGED 85 YEARS

24 TO THE

MEMORY OF
ANN THE WIFE OF
SAMUEL ACLAND
OF WILKES STREET
WHO DEPARTED THIS LIFE
JULY 25th ANNO DOMINI 1833
AGED 60 YEARS
ALSO THE ABOVE NAMED
SAMUEL ACLAND
WHO DEPARTED THIS LIFE
NOVEMBER 25th ANNO DOMINI 1844
AGED 69 YEARS

25* THE FAMILY VAULT
OF THE LATE
MR JAMES SIMPSON
28th FEBRUARY
1843

26 THIS PUBLIC VAULT BEING FULL WAS CLOSED IN THE YEAR 1845
REV. WILLIAM STONE MA RECTOR WILLIAM HICKS HALL
THOMAS MIDWINTER CHURCHWARDENS

27 CHRIST CHURCH MIDDLESEX
BY AN ORDER OF HER MAJESTY IN COUNCIL
DATED THE 2nd FEBRUARY 1867 IT WAS DIRECTED THAT
ALL THE COFFINS UNENCLOSED BENEATH THIS CHURCH
BE EMBEDDED IN SOIL MINGLED WITH CHARCOAL AND ENCLOSED
EITHER BY CONCRETE OR BY STONEWORK OR BRICKWORK
PROPERLY CEMENTED THE SEVERAL WORKS REQUIRED
BY SUCH ORDER HAVE BEEN PERFORMED
AND THE CRYPT AS FAR AS POSSIBLE CLEARED AND IMPROVED
UNDER THE DIRECTION OF
ROBERT ARNOLD, JOHN LIDDLE MEDICAL OFFICER OF HEALTH
BENJAMIN BROOKMAN

HENRY WOOLVEN CHURCHWARDENS
SOMERSET J. HYAM VESTRY CLERK
JAMES TOLLEY ARCHITECT 30th NOVEMBER 1867 T.C. TOLLEY BUILDER

28 W ...
THE UPPER VAULT NO 15
LIES THE REMAINS OF
MISS CHARLOTTE VAUX ...
JOHN VAUX WHO DIED NOV ...
AGED 90 [?]
... TWO INFANT SONS ...
ALSO MISS JANE ...
WHO DIED NOV ...

JOHN ...
ANN ...
MISS ...

29 LESCHALLAS

Appendix D

Details of coffin furniture design types from Christ Church, Spitalfields, giving number of dated examples/number retrieved/expressed as percentage, and date range (see microfiche M2–M3 for illustrations)

1 Upholstery pins

Type	Nos: dated/retrieved/%	Date range
01	046/104/44.2	1739–1843
02	016/035/45.7	1747–1839
03	020/047/42.6	1744–1833
04	008/017/47.1	1743–1821
05	000/002/00.0	
06	001/006/16.6	1821
07	000/004/00.0	
08	001/003/33.3	1792
09	010/023/43.5	1760–1825
10	002/003/66.6	1761–1849
11	005/021/23.8	1746–1811
12	001/002/50.0	1781
13	000/001/00.0	
14	005/010/50.0	1759–1825
15	001/002/50.0	1822
16	001/002/50.0	1754
17	000/001/00.0	
18	001/002/50.0	1809
19	000/001/00.0	
20	002/002/100	1752–1757
21	000/001/000	
22	001/001/100	1808
23	001/001/100	1813
24	009/019/47.4	1812–1852
25	001/005/20.0	1847
26	002/002/100	1815–1817
27	002/002/100	1750–1816
28	000/001/00.0	
29	000/001/00.0	
30	001/001/100	1813
31	001/002/50.0	1757
32	000/002/00.0	
33	000/001/00.0	
34	001/003/33.3	1823
35	001/004/25.0	1825
36	001/004/25.0	1842
37	001/001/100	1799
38	002/002/100	1802–1821
39	000/001/00.0	
40	004/007/57.2	1825–1839
41	000/001/00.0	
42	000/001/00.0	
43	000/001/00.0	
44	002/002/100	1819
45	002/002/100	1809–1826
46	000/002/00.0	
47	001/002/50.0	1820
48	002/002/100	1821–1839
49	000/001/00.0	
50	000/001/00.0	
51	000/001/00.0	
52	003/004/75.0	1770–1782
53	001/001/100	1794
54	000/001/00.0	
55	000/001/00.0	
56	000/004/00.0	
57	000/001/00.0	
58	000/004/00.0	

2 Grips

Type	Nos: dated/retrieved/%	Date range
01	006/029/20.7	1747–1847
02	026/088/29.5	1763–1837
03	025/121/20.7	1729–1827
04	059/176/33.5	1743–1847
05	017/072/23.6	1744–1835
06	007/019/36.8	1839–1849
07	002/002/100	1821–1849
08	000/001/00.0	
09	001/002/50.0	1770
10	001/002/50.0	1837

11	000/001/00.0	
12	000/001/00.0	

3 Grip plates

Type	Nos: dated/retrieved/%	Date range
01	003/005/60.0	1812–1825
02	000/001/00.0	
03	037/100/37.0	1768–1842
04	000/002/00.0	
05	009/015/60.0	1729–1815
06	000/001/00.0	
07	003/005/60.0	1791–1813
08	000/002/00.0	
09	009/022/40.9	1784–1827
10	000/001/00.0	
11	002/002/100	1795–1849
12	001/001/100	1761
13	001/001/100	1798
14	002/004/50.0	1843–1845
15	000/001/00.0	
16	000/002/00.0	
17	002/002/100	1765–1793
18	000/001/00.0	
19	001/002/50.0	1763
20	000/004/00.0	
21	000/001/00.0	
22	000/001/00.0	
23	000/001/00.0	
24	002/004/50.0	1794–1806
25	005/010/50.0	1833–1847
26	001/002/50.0	1819
27	001/002/50.0	1779
28	000/001/00.0	
29	001/001/100	1776
30	001/001/100	1747
31	001/003/33.3	1823
32	000/001/00.0	
33	004/008/50.0	1806–1828
34	001/001/100	1799
35	000/001/00.0	

4 Breastplates

Type	Nos: dated/retrieved/%	Date range
01	009/015/60.0	1729–1807
02	002/002/100	1839–1845
03	006/011/54.5	1810–1821
04	005/005/100	1783–1822
05	003/003/100	1827–1847
06	025/025/100	1783–1852
07	002/002/100	1779–1794
08	034/034/100	1767–1825
09	012/012/100	1773–1797
10	000/001/00.0	
11	000/001/00.0	
12	002/002/100	1788–1802
13	001/001/100	1799
14	004/004/100	1743–1818
15	002/002/100	1824
16	001/001/100	1835
17	001/001/100	1828
18	001/010/100	1765
19	001/001/100	1761
20	(Inscription type 005)	
21	021/021/100	1824–1847
22	001/001/100	1821
23	001/001/100	1831
24	004/004/100	1782–1819
25	001/001/100	1832
26	003/003/100	1834–1849
27	003/003/100	1788–1839
28	004/004/100	1829–1842
29	000/002/00.0	
30	003/003/100	1809–1832
31	003/003/100	1759–1821
32	001/001/100	1830
33	001/001/100	1802
34	001/001/100	1820
35	006/006/100	1806–1825
36	001/001/100	1821
37	001/001/100	1796
38	005/006/83.3	1779–1825
39	001/002/50.0	1794
40	001/001/100	1788
41	003/003/100	1764–1767
42	001/001/100	1777
43	002/002/100	1793–1797
44	002/002/100	1828–1829
45	000/001/00.0	
46	006/006/100	1771–1821
47	(Inscription type 008)	
48	001/001/100	1835
49	000/001/00.0	
50	004/006/66.6	1780–1821
51	001/001/100	1795
52	004/004/100	1778–1794

53	001/001/100	1834
54	001/001/100	1827
55	003/003/100	1820–1826
56	001/002/50.0	1825
57	002/002/100	1812–1824
58	001/001/100	1823
59	001/001/100	1793
60	000/001/00.0	
61	002/003/66.6	1765–1786
62	001/001/100	1811
63	001/001/100	1775
64	004/004/100	1777–1794
65	001/001/100	1778
66	006/006/100	1761–1770
67	003/003/100	1769–1777
68	001/001/100	1768
69	003/003/100	1765–1803
70	002/002/100	1777–1778
71	001/001/100	1765
72	001/001/100	1765
73	001/001/100	1776
74	001/001/100	1777
75	002/002/100	1782
76	002/002/100	1785–1793
77	001/001/100	1823
78	001/001/100	1827
79	001/001/100	1790
80	002/002/100	1777–1786
81	001/001/100	1836
82	005/005/100	1820–1829
83	001/001/100	1747
84	002/002/100	1833–1836
85	001/001/100	1835
86	002/002/100	1795–1811
87	001/001/100	1827
88	001/001/100	1770
89	001/002/50.0	1758
90	001/001/100	1827
91	001/001/100	1824
92	001/001/100	1832
93	001/001/100	1852
94	001/001/100	1829
95	002/002/100	1737–1746
96	001/001/100	1732
97	001/001/100	1793
98	001/001/100	1776
99	001/001/100	1772
100	001/001/100	1775
101	001/001/100	1768
102	001/001/100	1739
103	002/002/100	1806–1809
104	002/002/100	1784–1789
105	001/001/100	1753
106	000/001/00.0	
107	001/001/100	1794
108	001/001/100	1806
109	000/001/00.0	
110	001/001/100	1827
111	001/001/100	1788
112	001/002/50.0	1757
113	001/001/100	1811
114	001/001/100	

5 Escutcheons

Type	Nos: dated/retrieved/%	Date range
01	015/045/33.3	1776–1827
02	001/002/50.0	1839
03	001/006/16.6	1815
04	010/024/41.6	1779–1839
05	000/003/00.0	
06	004/010/40.0	1823–1835
07	000/001/00.0	
08	000/001/00.0	
09	001/005/20.0	1779
10	004/017/23.5	1779–1839
11	002/004/50.0	1832–1845
12	012/030/40.0	1779–1847
13	002/011/18.2	1833–1835
14	005/007/71.4	1811–1822
15	000/001/00.0	
16	001/002/50.0	1842
17	000/001/00.0	
18	000/001/00.0	
19	000/002/00.0	
20	000/001/00.0	

6 Lid motifs

Type	Nos: dated/retrieved/%	Date range
01	001/005/20.0	1839
02	016/039/41.0	1795–1847
03	004/010/40.0	1821–1824
04	000/006/00.0	
05	001/002/50.0	1798
06	011/030/36.6	1779–1847

07	001/001/100	1849
08	003/003/100	1832–1849
09	001/001/100	1849
10	002/003/66.6	1793–1820
11	002/005/40.0	1822–1843
12	000/001/00.0	
13	000/003/00.0	
14	000/001/00.0	
15	000/001/00.0	
16	001/001/100	1789
17	002/002/100	1821–1824
18	000/001/00.0	
19	000/001/00.0	
20	000/001/00.0	
21	000/001/00.0	
22	001/001/100	1794
23	000/002/00.0	
24	001/001/100	1798
25	000/001/00.0	
26	000/001/00.0	

7 Inscriptions

Type	Nos:dated/ retrieved/%	Date range	Shape
001	054/054/100	1739–1843	Rectagular plain
002	013/013/100	1732–1831	Rectangular plain border
003	005/005/100	1816–1828	Directly on coffin
004	054/054/100	1784–1847	Directly and border
005	003/003/100	1813–1847	Trapezoidal plain
006	003/003/100	1822–1852	Trapezoidal plain border
007	001/001/100	1765	Lozenge plain
008	001/001/100	1828	Shield plain

Appendix E

Metallurgical analysis of a selected sample of coffin furniture

Paul Budd, University of Bradford

> Nothing can be more hideous, than the raised metal work called coffin furniture, that is so generally used at the present time; heathen emblems, posturing angels, trumpets, death's heads and cross bones, are mingled together in a glorious confusion, and many of them partake of a ludicrous character.
>
> (Pugin 1844)

A total of 64 items of metal coffin furniture, recently excavated from the vaults beneath Christ Church, Spitalfields, London, were examined. The sample was taken from a total of about 800 pieces of coffin furniture recovered during the excavation and was selected to be representative of the range of metal types and manufacturing techniques noted. Samples from all the coffin furniture examined were analysed qualitatively by energy dispersive X-ray fluorescence (XRF) in order to identify the base metals and coatings used. All of the coffin furnishings were then examined visually and under a low-power microscope in order to determine some details of the technique of manufacture.

The coffin furniture which was provided for analysis was identified first by its individual context number and second by the terms breastplate, inscription, lid motif, grip plate, grip, and escutcheon. These terms had been taken directly from contemporary pattern books of the period and had been used throughout the excavation. The coffin furniture may be further divided according to the base material used in its construction.

Four base materials were used in the manufacture of the coffin furniture: brass, lead, iron, and tinplate (tin-coated iron sheet). It is notable in this context that significant signals for tin were detected on much of the coffin furniture described by the excavators as iron. The tin signals suggest that these artefacts are in fact badly corroded tinplate. Although there were no visible traces of tin on these examples, this should not be considered too surprising given the state of preservation of the tin layers on some of the coffin furniture known to be tinplate. Very thin coatings could be achieved by the early 19th century. The following is a brief summary of the results of the examination and analyses of the coffin furniture.

The breastplates varied in size, but were generally the largest items of coffin furniture. They were usually rectangular in form, although some oval and diamond-shaped examples were noted. Trapezoidal and other less geometric shapes have also been recorded. All the breastplates examined were inscribed with brief biographical details and almost all were decorated in some way (the exceptions being some of the brass plates which contained the lettering of the inscription only). In all cases the decoration left a blank central field to which the lettering had been added. XRF results indicate that the breastplates examined were made from one of three base materials: brass, lead, or tinplate. Three of the four brass breastplates examined were undecorated, containing only the lettering. These were made from fairly thick (typically 1–2 mm) sheet, probably cut from a larger sheet, although no tooling marks were apparent on the sides on the plates. None of these plates display any traces of surface coatings and they would presumably have been polished. Examination of tool marks within the lettering suggest that they were hand engraved. The remains of black paint were noted within the engraving on two of the plates, and clearly this was intended to highlight the wording. On one of the plates the engraving was inlaid with a black pigment to acheive a similar effect. XRF analyses of these paints and the pigment failed to detect any significant signals, suggesting that they are possibly organic. The remaining brass breastplate featured a wide, decorated border as well as lettering in the central field. The plate was formed from thin sheet metal and the decoration was raised by repeated striking with a small circular punch. The plate was too badly corroded to allow examination of the tooling marks with the lettering, although the plate would have been too thin to allow engraving and the lettering would probably have been chased (ie impressing the metal to form a groove by hammering a hand-held tool) on to the blank field.

The lead breastplates fell into two categories. Some examples were generally more massive with a less uniform thickness. They showed little detail of the decoration on the reverse, although the reverse side generally followed the contours of the relief decoration. This suggests that they were cast directly. Examination of the decoration suggested that in all cases the decoration had been 'touched up' to some extent after casting by hand engraving. Some of the breastplates also display punch marks on their reverse in areas of the highest relief decoration, indicating that a punch was used in some cases to raise the metal and correct imperfections in the design 'as cast'. The remaining lead breastplates were thinner and showed detailed patterning on the reverse side of the plate, indicating that they were pressed from lead sheet. The lead breastplates were 1.5–4 mm thick, and there is clear evidence from the tooling marks that the lettering was engraved in all cases. Four of the six plates examined showed signs of having been coated: one in silver, one in brass, and two in black lacquer or paint which could not be identified by XRF and may be organic.

The tinplate breatplates examined were typically less than 0.5 mm thick and featured embossed decoration. Most were clearly manufactured by pressing from sheet metal, although on one (2166) the design was probably raised using a small punch, and at least one other was too badly corroded to be certain of the technique. Numerous examples were found where clear tool marks could be seen with the lettering of the inscription, and in all but one case the technique used was chasing. The only exception is where a letter punch has been used. Most of these plates displayed traces of black lacquer or paint. This was applied only to the top surface of the plate; splashes and finger marks on the reverse side suggest that it was painted on by hand. XRF analysis of the lacquer gave no significant signals, and again it is possible that the material is organic.

Four inscriptions were examined, all of which were from lead coffins. Either a rectangular piece of lead was cut from a sheet, inscribed with biographical details and then soldered on to the coffin, or the biographical details were inscribed directly on to the coffin after it was sealed. Clear tool marks show that all of the inscriptions were made by engraving. On two of the inscriptions a simple decoration had been provided by repeated striking with a decorative punch to form a rectangular border of small circular motifs.

The lid motifs and escutcheons are discussed together since they are essentially the same, being decorative embellishments from different parts of the coffin. All the lid motifs and escutcheons examined featured embossed designs and were pressed from thin sheet metal, details of their decoration being clearly visible on their reverse sides. The majority of the lid motifs examined were tinplate, as were seven of the eight escutcheons. Many of these were black-lacquered or painted on the top side. XRF of the lacquer gave similar results to the analyses for those from the tinplate breastplates, and it is likely that the material is the same in both cases. One of the escutcheons examined and one of the lid motifs were brass. Two of the lid motifs were lead.

Most of the grip plates were of very similar construction to the other coffin furniture, all but two being pressed from sheet metal to form an embossed design. Again tinplate was the most common material, and again numerous examples were black-painted or lacquered. Two of the grip plates differed in their bulkier appearance and lack of design detail on the reverse side, and these were almost certainly cast. One of these is lead, coated with brass. The other is iron and shows no sign of a surface coating, although it is very corroded and it may have originally been painted.

All the grips examined were solid metal. A few had casting flashes clearly visible, running lengthwise along the grip, suggesting that they were cast in two-piece moulds. One of the grips was brass; all the others were iron. Two of the iron grips still had traces of black paint attached. The other iron grips were badly corroded and may also have originally been painted. No traces of tin were detected on any of the grips.

The manufacture of the coffin furniture may be broken down into a number of basic processes, which are discussed in detail below. Where the coffin furniture was made from sheet metal or tinplate the first of these processes was the manufacture of the base material. This was followed by a sheet metal forming process in order to produce the embossed decoration. Biographical details were then inscribed where appropriate. Where the coffin furniture was cast directly, a single fabrication process was used to produce the artefact with its decoration, although subsequent hand finishing was necessary. The final process in many cases was the addition of lacquer or coating.

Tylecote (1976, 131) has described the manufacture of brass sheet in the 18th century by pouring the molten metal into horizontal stone moulds. The resulting plates were then hammered into sheet. Copper was being rolled into sheet in the 17th century (Raistrick 1972, 38), and rolling mills strong enough to roll brass were certainly being built by the end of the 18th century (Garanger 1968). It is probable that the sheet metal used in the manufacture of all of the brass coffin furniture was rolled, although it is only possible to observe details of the original metal sheet on the three undecorated breastplates (2263, 2259, and 2486), since these have not undergone a subsequent fabrication process. These three breastplates are of uniform thickness and have a smooth surface with no evidence of hammer marks. One of them (2259) shows shallow linear striations running the length of the plate, which may have been caused by defects in the rollers.

Metallographic examination of a small sample from the breastplate 2259 revealed equiazed polygonal grains with some twinning, a structure characteristic of a fully annealed brass. This suggests that the final process involved annealing the brass (ie heating it above its recrystallization temperature, typically 400–700 degrees C). It is possible therefore that the plate was produced by hot rolling. However, sheet metal is normally rolled cold in order to obtain a good surface finish. This can lead to cracking if the metal is not softened by annealing between rolling passes. It is most likely that the manufacturing process in this case was one of alternate cold rolling and annealing, ending with a final anneal to leave the metal in a soft condition suitable for engraving.

The manufacture of lead sheet by casting lead on to a flat bed with clay-lined wooden sides is well known (Tylecote 1986, 76). By the later 18th century sheet cast in this way was being challenged by rolled products (Rowe 1983, 8). Raistrick (1972, 37) has dated the introduction of lead rolling into Britain to 1670. The manufacture of tinplate in the 18th and 19th centuries has been described by Smith in his review of the British tinplate industry (Smith 1981–2). By the mid 18th century a flourishing tinplate industry had developed in Britain based on rolled wrought iron as a base material. Bar iron was repeatedly hot rolled, folded (or 'matched') and then rerolled in 'pairs' and then 'fours' and finally into 'eights'. The folds were then trimmed off and the plates separated. After rolling the plates were 'pickled' in an acid solution and cleaned abrasively to remove scale and corrosion and to prepare the surface for coating. In the tinning process plates were first dipped in hot grease to increase the flow properties of the tin and then lowered vertically into the molten tin (which was held under grease to prevent oxidation). From the mid 19th century mechanical rollers were introduced to feed plates into the tin pot, allowing more even coatings to be produced without the imperfections caused by handling with tongs.

The majority of the coffin furniture examined had been embellished with decorative motifs. The only exceptions were some of the brass breastplates and lead inscriptions which consisted only of engraved lettering on an otherwise plain sheet of metal. All these decorations were created by fabricating the base metal. The designs are of varying complexity, but in almost all cases the decoration takes the form of a relief pattern. Three different fabrication techniques have been noted: casting, punching, and pressing.

The simplest fabrication technique was to cast the coffin furniture directly. The cast coffin furniture is generally bulkier and stronger than that manufactured by other techniques. The most commonly cast metals were lead and iron, although one brass casting was also noted. The technique was most commonly used to manufacture grips; indeed, all the grips examined had been cast (this would have been the easiest way to manufacture objects of this shape). The manufacture of cast lead sheet has been outlined above, and Tylecote (1986, 75) also describes the casting in of decorations by imprinting a pattern into the sand bed prior to pouring the molten lead. This technique was used in the manufacture of 17th and 18th century cast iron graveslabs and firebacks (Willats 1987). Much of the detail on the cast lead coffin furniture appears to have been added by hand punching and engraving after casting, and it seems likely that the 'as cast' product was not regarded as finished.

Repeated striking with a small hand-held punch to form a raised design was noted on one of the brass breastplates (2675), and was possibly a technique used for other decorated thin sheet metal coffin furniture such as breastplate 2166. There were two other examples of punching. These were the two inscriptions where repeated striking with a fairly large patterned punch was the technique used to provide a decorative border. A hand-held punch of this size would have been used to best effect only on a relatively soft metal and it is notable that in both cases the base material was lead sheet.

Church and Smith (1966) refer to the invention, by John Pickering of London in 1769, of a mechanical stamp or press for raising a pattern in coffin furniture and other brass foundry. This replaced the labour-intensive task of punching sheet metal to form a design and led to a significant decrease in the price of decorated coffin furniture. By far the most common fabrication technique was the mechanical pressing or stamping of sheet metal with a pattern (or 'die') in order to form an embossed design. This technique was used for lead, brass, and all the tinplate material, and possibly also for iron sheet. The technique allows a greater economy of metal than either undecorated engraved plates or cast coffin furniture since thin sheet metal can be used.

Lettering occurs on the breastplates and inscriptions. Three different techniques of lettering have been noted on the basis of tooling marks: engraving, chasing, and letter punching. Engraving was the most common and letter punching the least, being found on only one example. The technique of engraving involves using a sharp hand-held tool to cut away at a metal surface. Engraving is possible therefore only where sheet metal is thick enough to allow cutting. The engraving tool can be rocked from side to side as it is pushed through the metal, which produces a characteristic series of 'V'-shaped cuts within the groove (Lowery *et al* 1971). Such marks were clearly visible on some of the lead coffin furniture where the soft metal would make engraving particularly easy. Chasing involves moving across the metal surface with a tool (or 'tracer') held in one hand, which is driven at frequent intervals into the surface of the metal using a hammer held in the other. The key difference is that the metal is indented rather than cut away. This technique is more appropriate than engraving when thin sheet metal is being worked. Clearly it would not be possible to engrave tinplate. Examples (where diagnostic marks remained), with one exception, had been chased. The only exception was one breastplate where the lettering had been made up by driving a series of letter punches into the surface.

Much of the coffin furniture displayed traces of a final surface coating added after the decoration and lettering. By far the most common coating was a black lacquer or paint which gave no significant XRF signals and is probably organic. The lacquer is never found on the reverse of the furniture and was probably painted on to the surface of the finished plate by hand. The application of hard, glossy, black varnish is often referred to as 'japanning', although the term can be confusing since it is applied to a range of different coating materials. Two pieces of coffin furniture, both cast lead breastplates, had thin metal coatings, one of brass and one of silver. In neither case is there evidence of the technique used, although it is possible that the metal was applied as leaf.

In the majority of cases coffin furniture was nailed to coffins. Small holes were noted around the edges of many of the furnishings and in a number of cases small nails remained *in situ* on the artefact. With all the tinplate and lead coffin furniture, where nails remained *in situ* they were made of iron. However, brass nails had been used to secure at least one of the brass breastplates. The grips were secured by a different method. In almost all cases where the grips survived attached to the grip plate, the grip was secured by means of an iron ring at each end. The ring was made from a strip of iron which was folded back on itself and sharpened at the end to form a spike which was inserted through a hole in the grip plate and into the coffin wall, often being folded back on the inside of the coffin. It is interesting to note that the use of this system of attachment makes it unlikely that the grips could have supported a great weight. The only exception noted was one iron grip plate where the grip was secured to loops projecting from the plate. The plate was then screwed into the coffin wall.

Despite restricting the detailed examination of the metalwork to a relatively small number of pieces, some general comments can be made. It is clear that over the period of use of the crypt, whereas there is no major change in the type of materials used in the manufacture of coffin furniture, there is considerable technological development in manufacturing techniques.

The earliest coffin furnishings examined were simple engraved brass sheet without decoration. There is at least one early example of an embossed decoration on thin sheet metal, but the labour-intensive task of raising the design by hand punching must have been a clear disincentive for the use of such decoration. In this context it is not surprising to find such designs produced by the simpler process of casting, although it is probable that the cast coffin furniture required considerable hand punching and engraving to achieve a satisfactory finish. In general this early coffin furniture was probably produced by skilled craft workers using a range of simple hand tools. The later 18th century sees the almost complete replacement of these essentially hand-made items by mechanically embossed designs in sheet metal. This period also sees the rise in popularity of tinplate. This is almost invariably painted or lacquered and was therefore obviously not regarded as decorative. Tinplate was presumably favoured over iron because it was rust-proof and therefore more durable; it was probably used in the manufacture of coffin furniture as a cheap alternative to brass or lead. By the mid 19th century the majority of the coffin furnishings were clearly mass-produced products which were 'personalized' by the addition of an inscription with the minimum of specialist labour.

Appendix F

Summary of a report on the examination of a ferrous coffin (2247) from Christ Church, Spitalfields

Brian Leddington, University of Bradford

Description

The coffin (see Fig F1) belonged to an infant interred in the upper north tunnel (UN). It consists of 7 major parts: 2 sheet metal sides; 1 sheet metal bottom; 2 rolled strip stiffening ribs to side; 2 rolled and forged stiffening ribs to ends. All the parts were held together by the only joining technique used on the coffin — rivets. The coffin was constructed of folded and riveted sheet metal, and measures 0.6 m in length. In form it is of the single-break type and resembles the traditional 'coffin-shape', but it is the only example of an all-ferrous coffin recovered from Christ Church. The coffin was not recovered complete. It had no lid, but two dovetails cut into the stiffening ribs on either side of the coffin appear to represent the remains of a lid-locking mechanism, which would have been intended to deter 'resurrectionists' (Fig F2).The interior of the coffin bore signs of whitewashing, which had been handbrushed along the lengths of the riveted joints. Traces of black paint were also visible on the interior sides. Along the bottom outside edge of the coffin was a small fragment of metallic coffin lace trim (Fig F3).

The two outer sides and the one remaining end appear to have been covered with textile or paper, strongly adhered to the sides. This covering shows silhouettes of coffin furnishings which originally decorated the coffin but which have now disappeared (see Fig F4).

Metallographic procedure and results

From the right-hand shoulder of the coffin three small sections were taken using a fine-toothed junior hacksaw: a straight cross-section of the stiffening rib and side sheet, bisecting the joining rivet; a longitudinal section consisting of the same components taken through an adjacent rivet; and a longitudinal section taken from a section of the stiffening rib between the first two samples. The samples were deburred and the examination faces cleaned with a smooth cut file. They were then individually mounted in a quick-drying 'Acryfix' resin mould and prepared for examination by hand grinding on decreasing grades of 'wet and dry' papers and polishing on two polishing pads impregnated with a diamond compound of 6 micron and 1 micron.

Under a reflecting microscope an approximate slag content was established, using the point counting method of Rostoker & Dvorak (1977). From an averaging of several counts the following counts were produced:

Specimen 1 stiffening rib = 13.3%
Specimen 1 side sheet = 9.1%
Specimen 1 rivet = 11.5%
Specimen 2 stiffening rib = 12.7%
Specimen 2 side sheet = 7.3%
Specimen 2 rivet = 10.3%
Specimen 3 stiffening rib = 13.9%

From metallographic examinations of the samples in both etched and unetched states the following inferences may be made: that all the samples consist of a single-phase alpha iron with little evidence of carbon content and are therefore ferrite. There is, however, a fairly high phosphorous content (which tends to take precedence at the expense of carbon in irons), which shows as a ghosting in the samples. There is also a fairly high slag content in the iron which shows it to be fairly poor quality but perfectly satisfactory for this type of construction. In essence all of the component parts of the samples are low-grade phosphoric iron (see Fig F5a–g).

Closer examination of the slag bands gives clues as to manufacturing techniques: in specimen 1 (Fig F5a) the stiffening rib shows long, thin ribbons of slag inclusions which implies forming along its length from a rolled bar. Rolling was probably done hot since there was no evidence of grain deformation. The rivets were similarly made in the hot state but used cold during the manufacture ofthe coffin.

Fig F6 shows a cross-section through a stiffening rib and rivet. The stiffening rib shows considerable plastic deformation in the 'necked' or narrowest part, and there is a 45 degree countersink on the sheet side coresponding to the head of the rivet. The hole produced in the stiffening rib is larger than necessary for the passage of the rivet. From these observations it may be inferred that the holes for the rivets were either pierced or machine-punched, and not drilled or hand-punched. The countersink was apparently created during the process

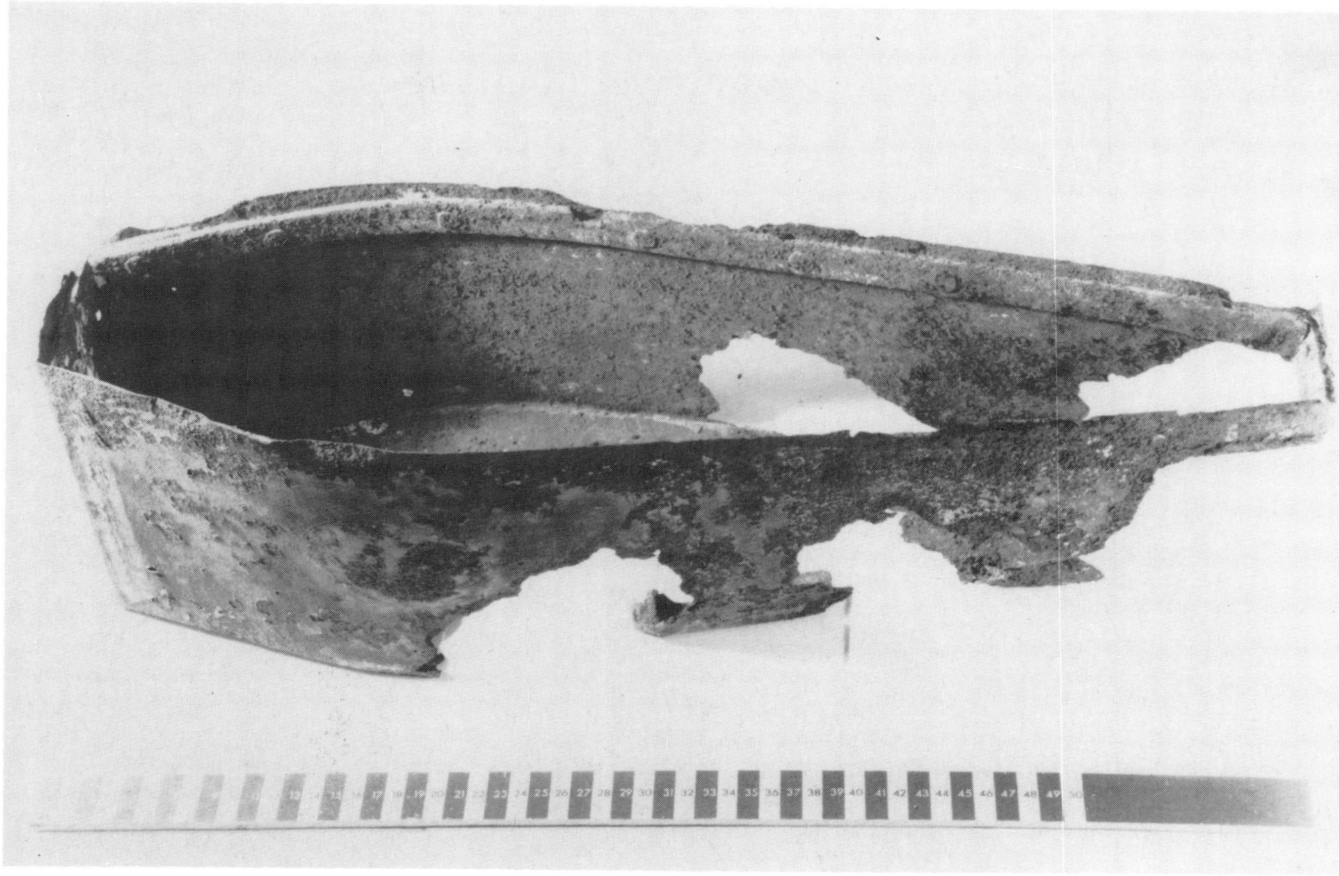

Figure F1 *The ferrous coffin after cleaning*

Figure F2 *Detail of dovetail used in locking device*

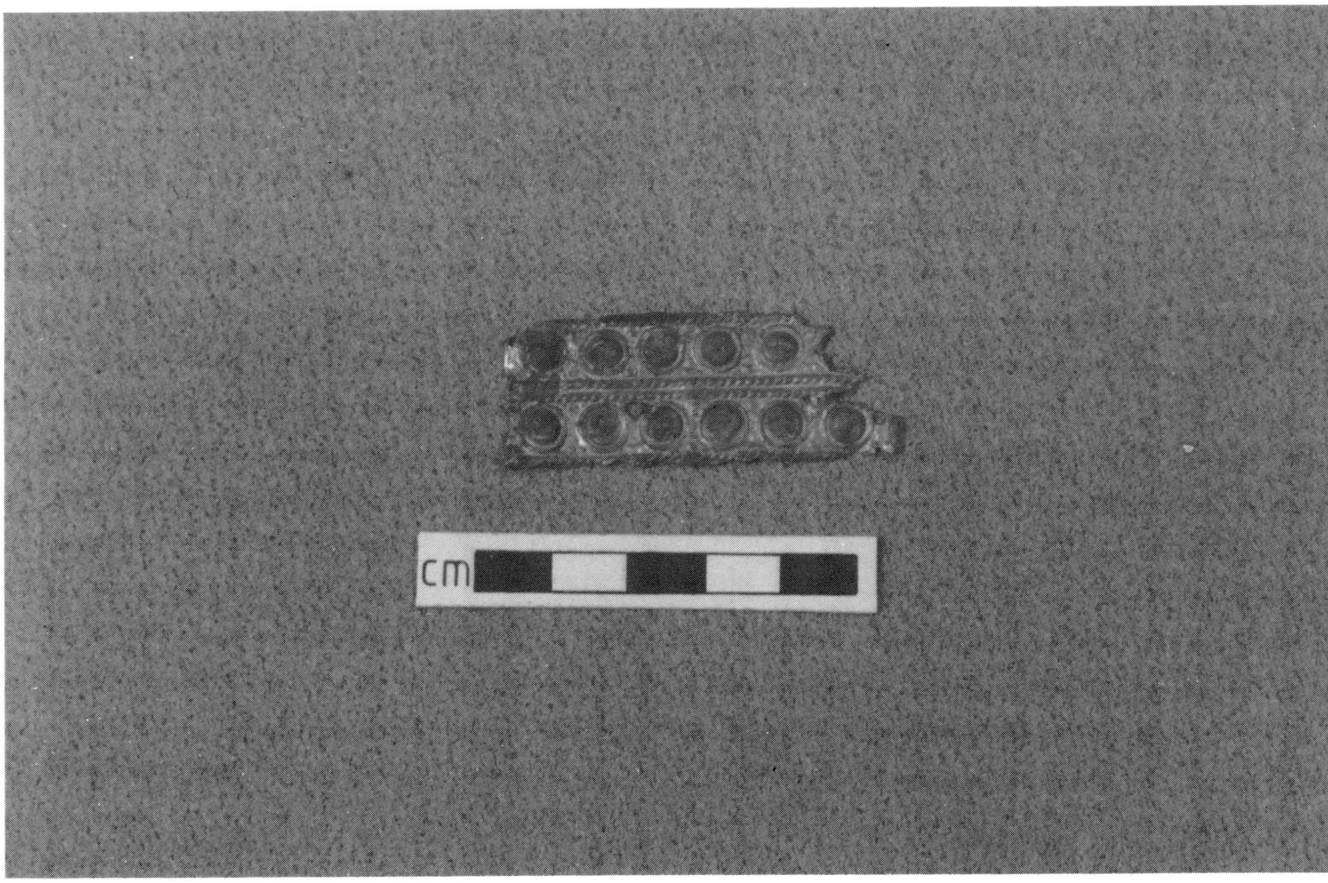

Figure F3 *Detail of metal lace decoration*

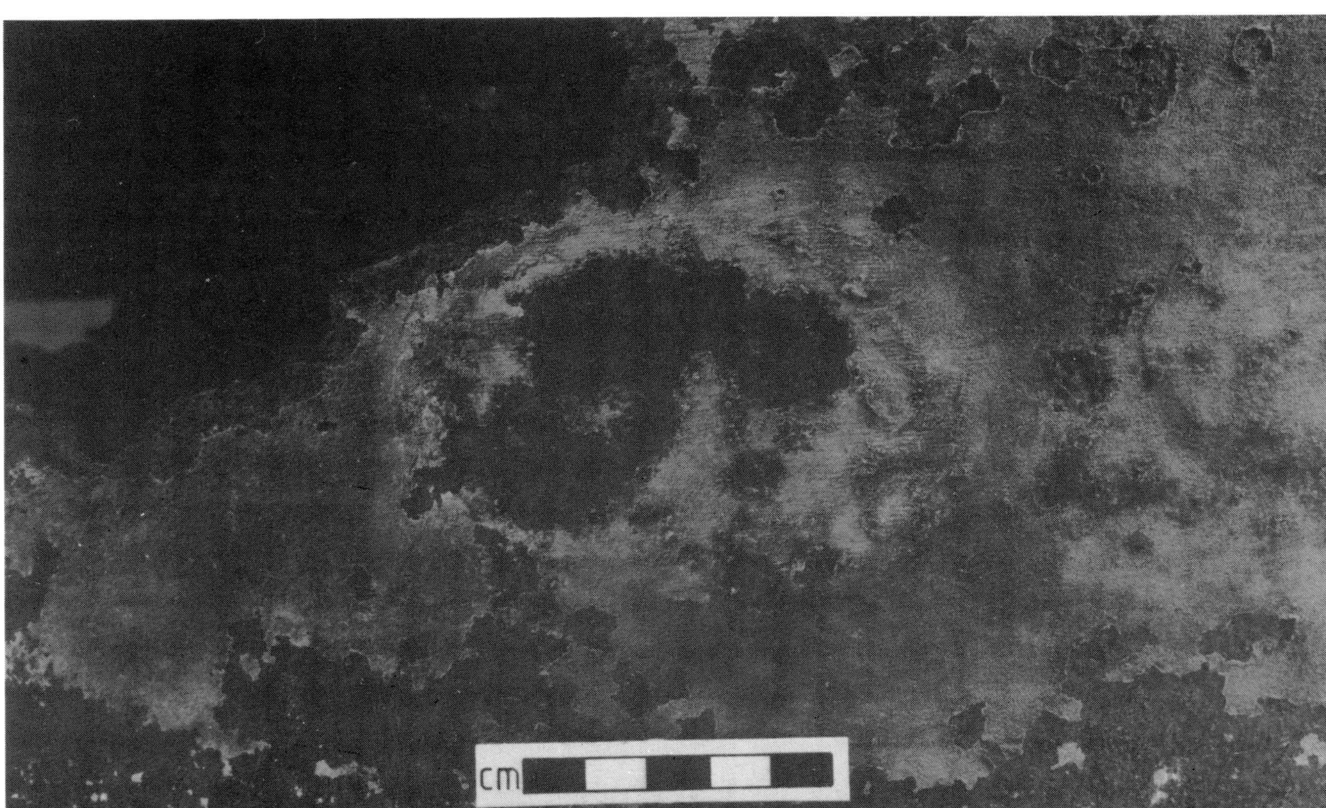

Figure F4 *Detail of coffin side showing silhouettes of furniture*

Figure F5 Surfaces of samples under reflecting microscope (a)

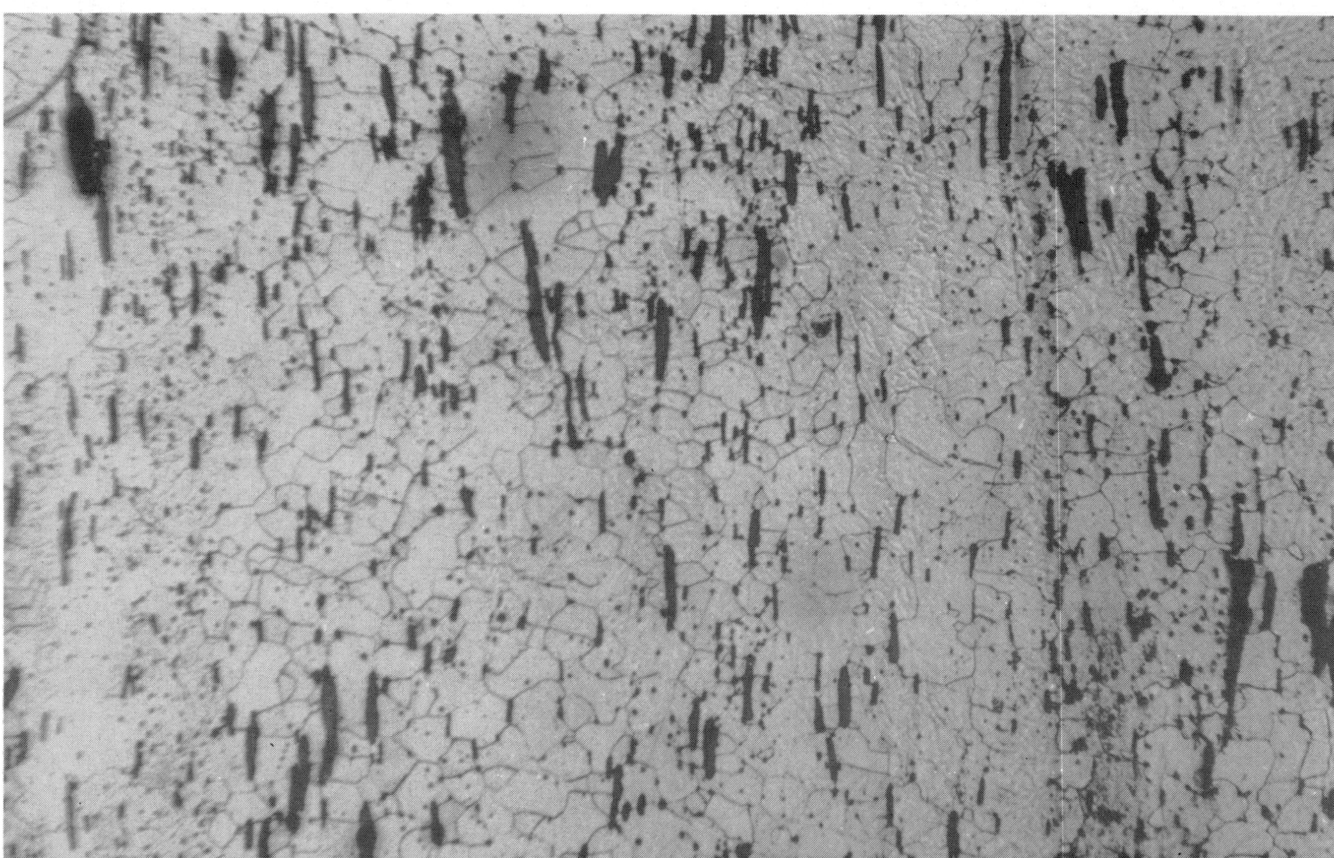

Figure F5 (b)

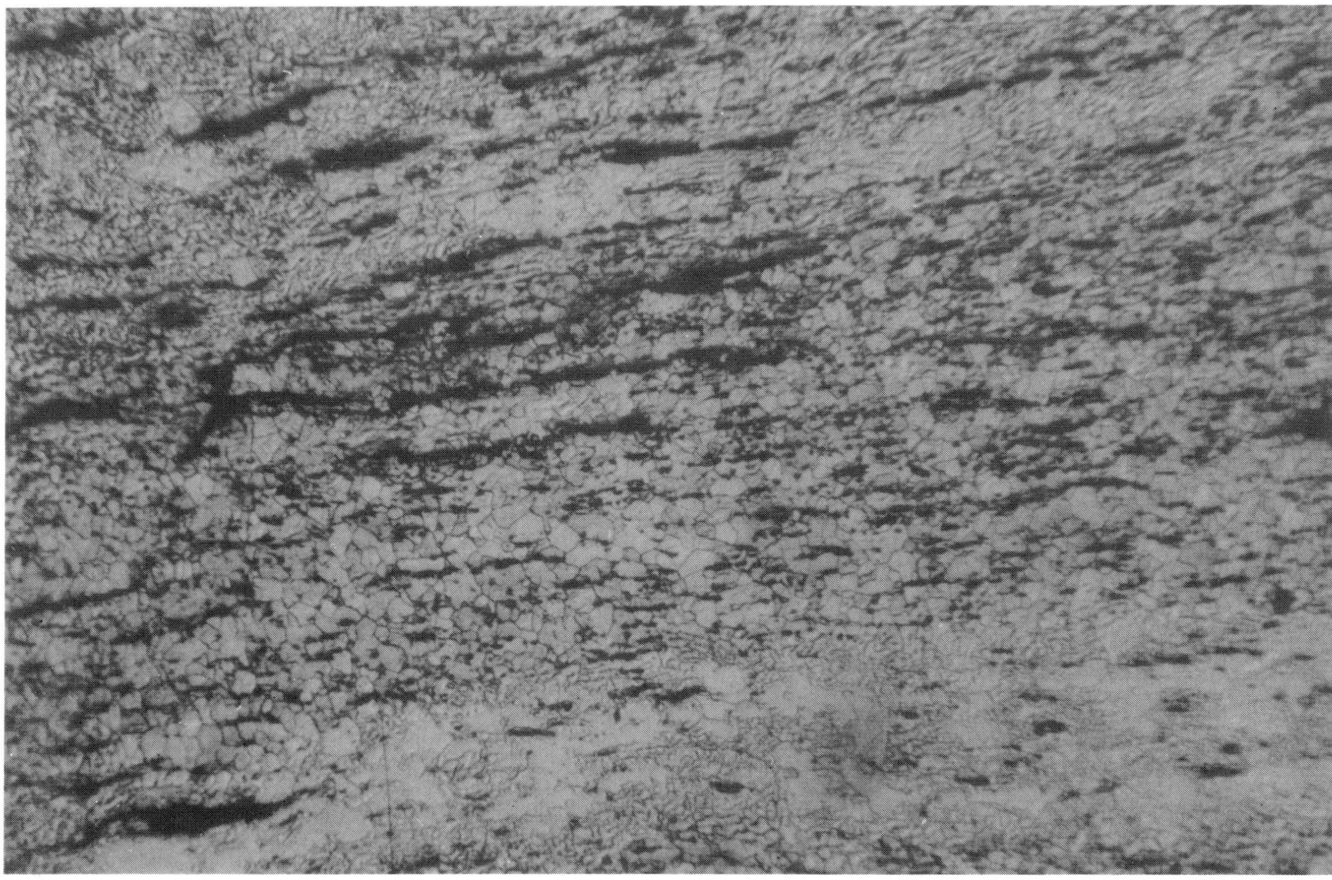

Figure F5 (c)

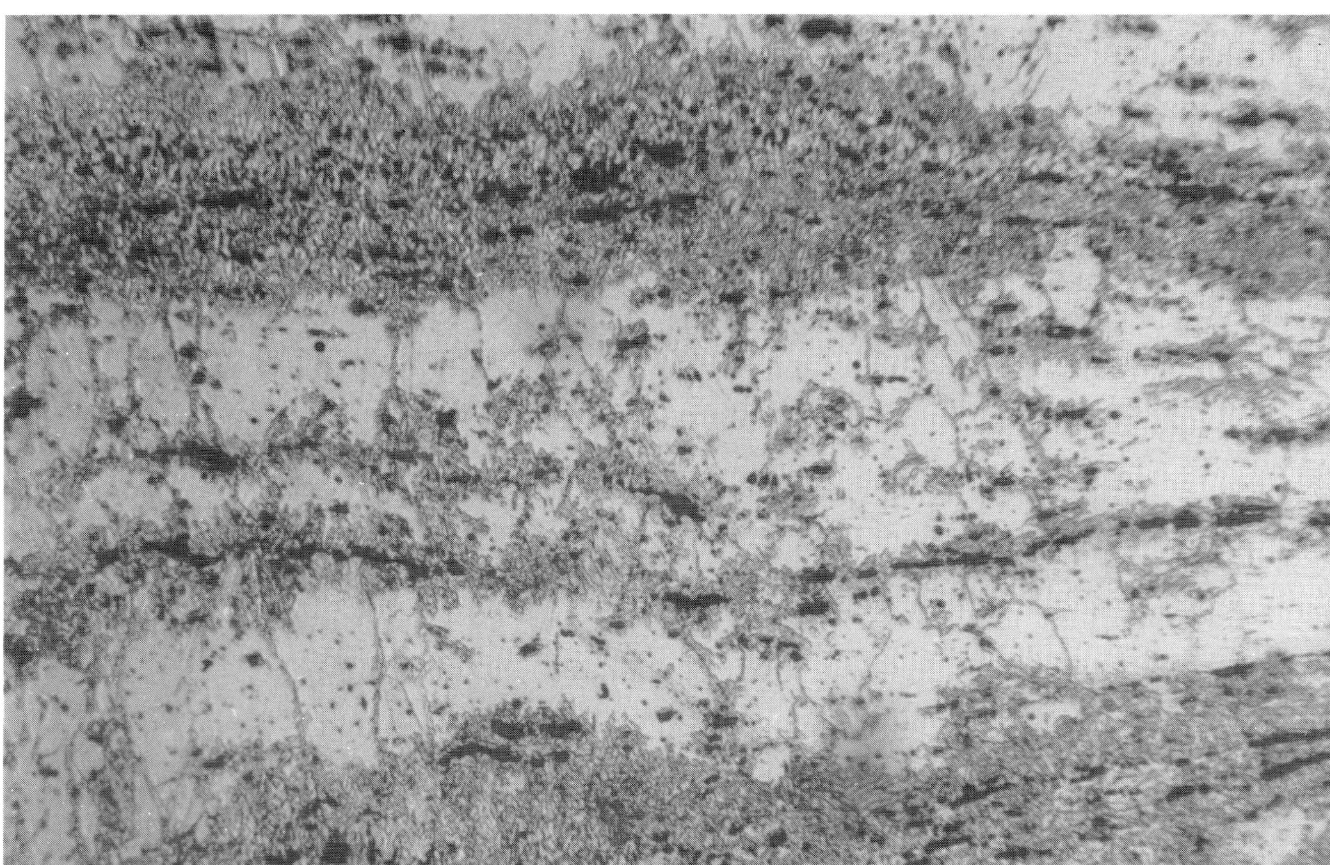

Figure F5 (d)

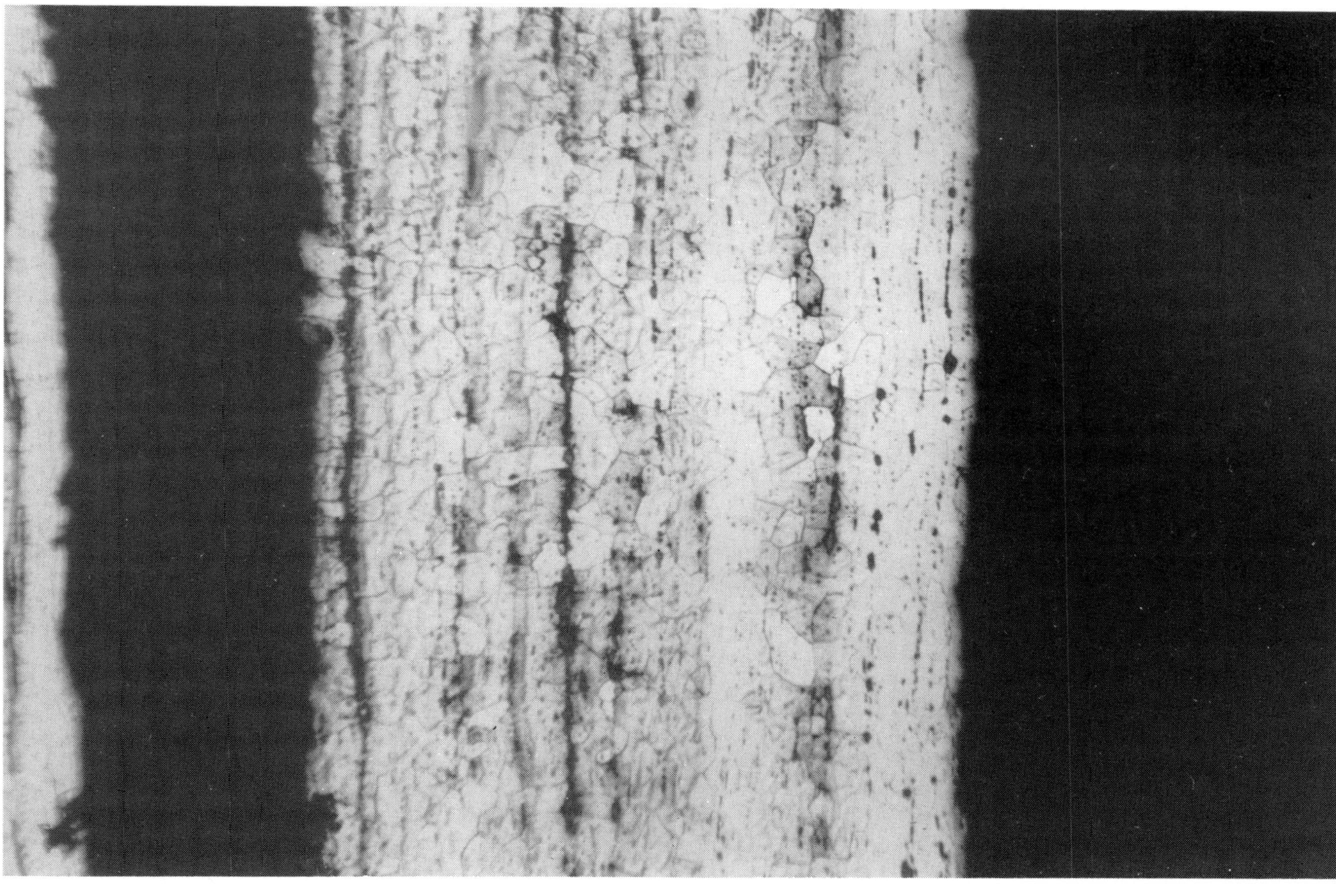

Figure F5 (e)

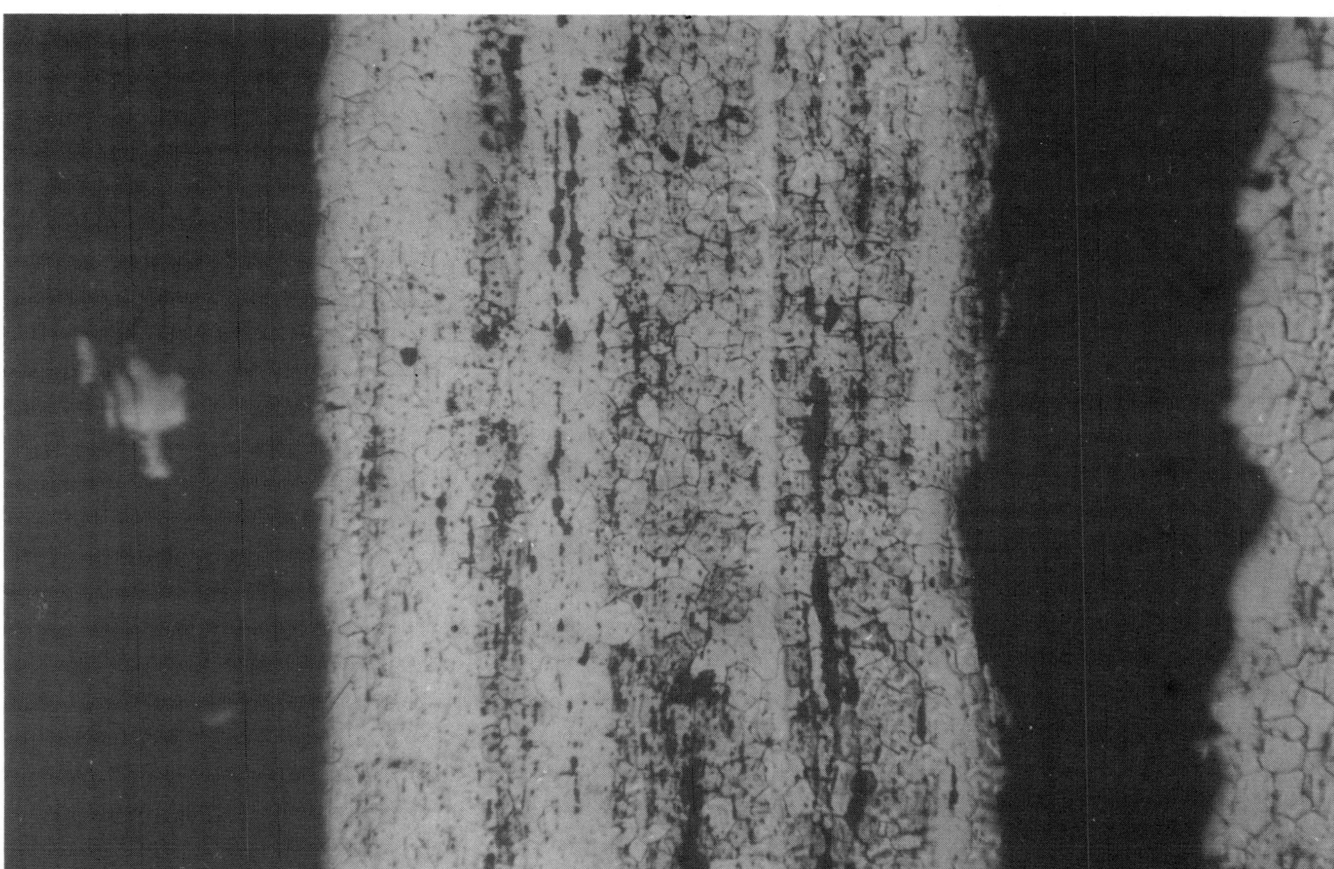

Figure F5 (f)

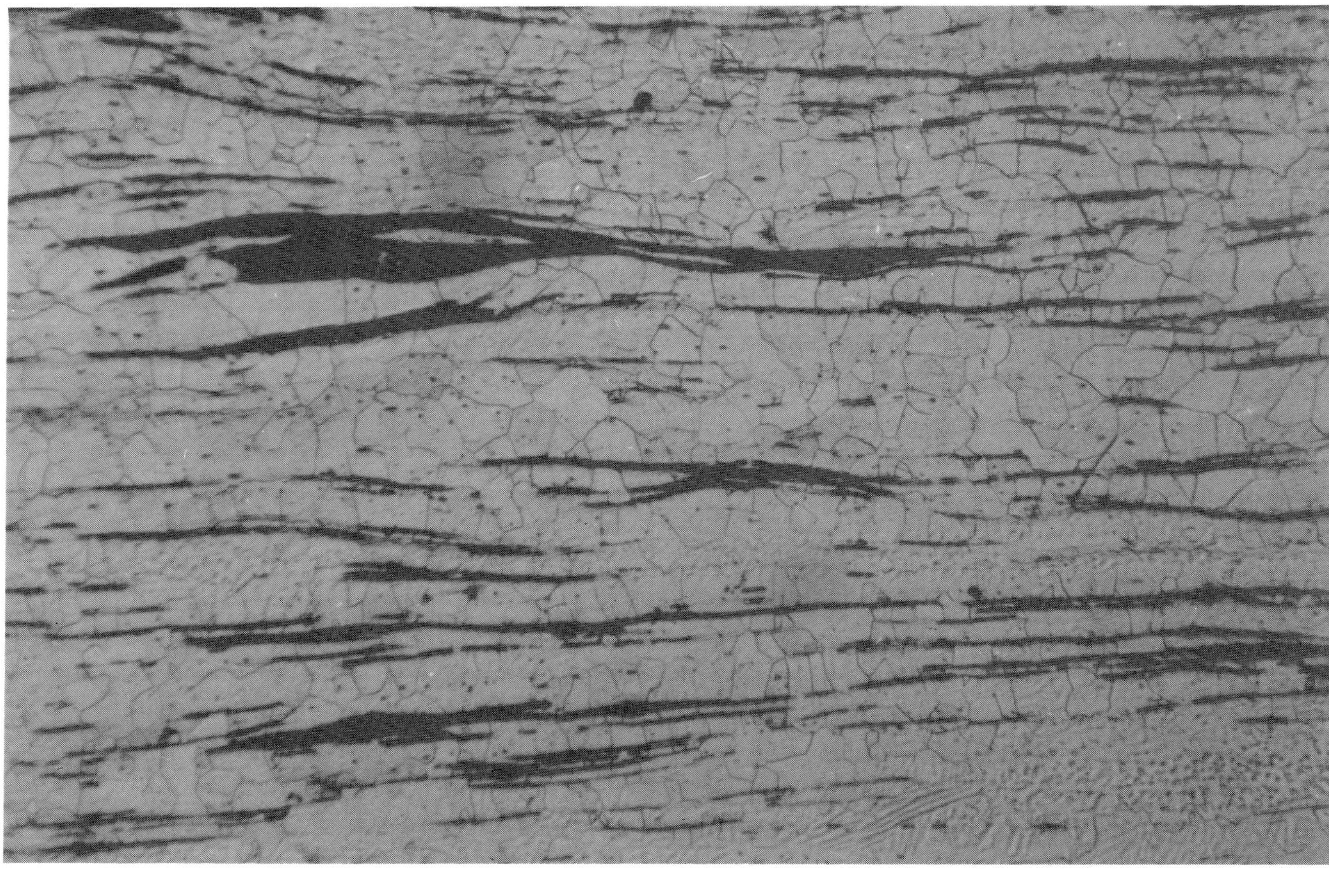

Figure F5 (g)

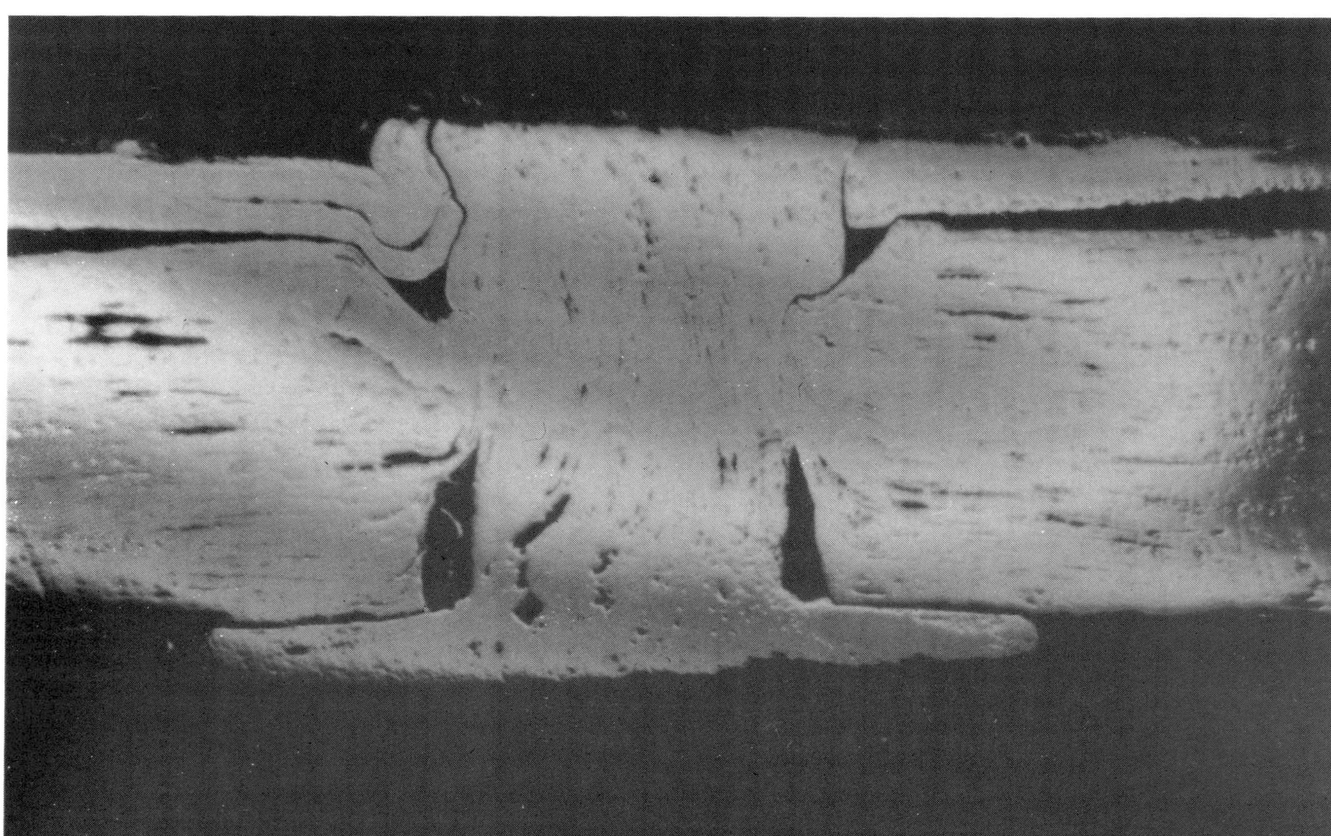

Figure F6 Cross-section of rivet

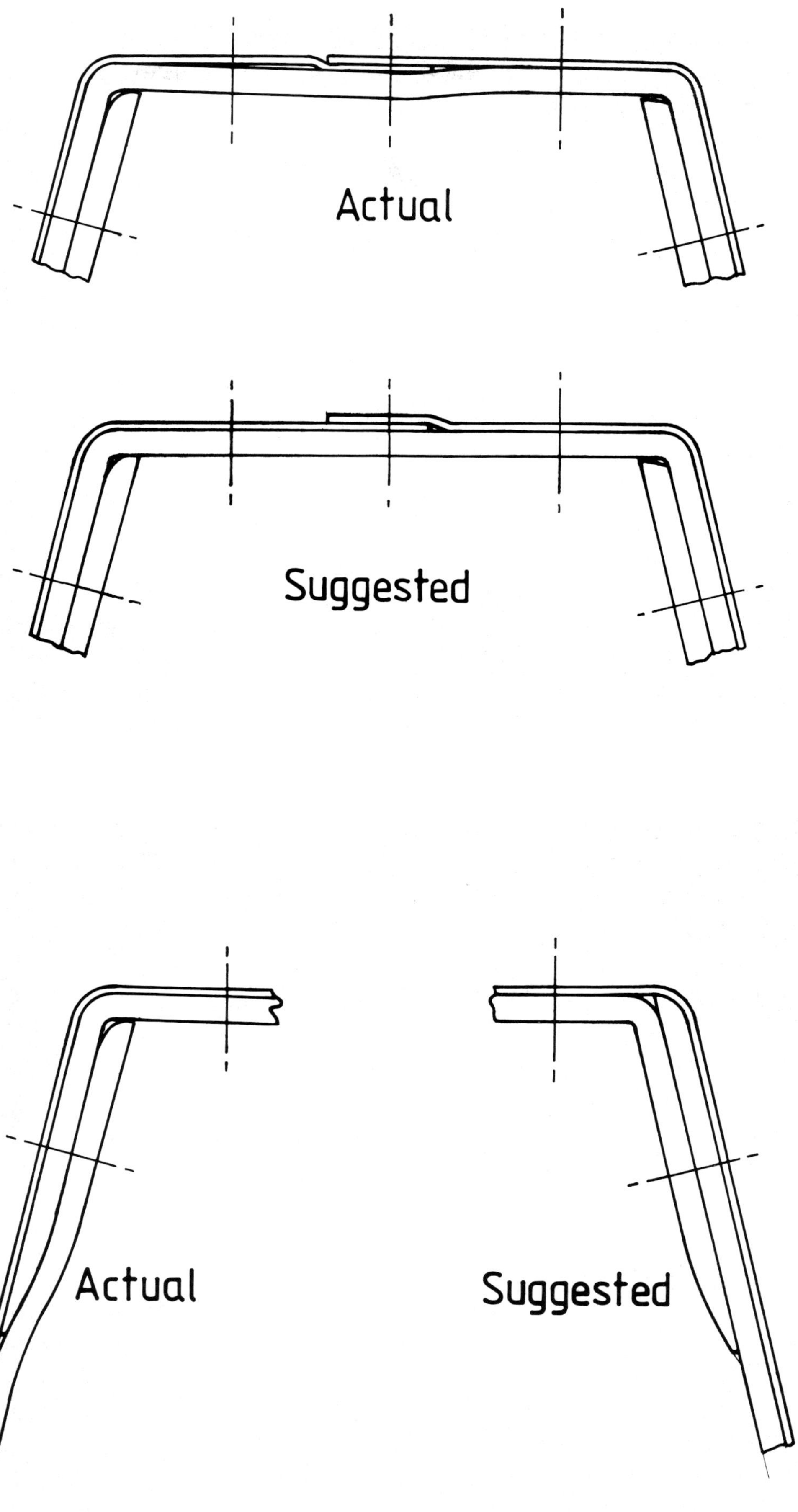

Figure F7 Details of actual and suggested joining methods (2247)

of pressing the rivet, which also accounts for the macrographic deformation of the sides of the hole. The oversize hole at the head end of the rivet was caused by the shape of the punch, which was apparently tapered. This would have been a poor tool for such a purpose since it would have been subject to considerable flank wear. The need for an accurate alignment of the pressing tool, together with evidence for a cold operation, implies that the hole was created by machine.

Manufacturing details and techniques

Evidence of forging is exhibited by the two end stiffening ribs and at the ends of the side stiffening ribs. Since phosphoric iron is very brittle in its cold state, the bending of the ribs to an angle of about 74 degrees and the piercing of the ribs by rivet holes probably required pre-heating of the bars. The stiffening rib at the foot end of the coffin has had to be specially forged to form the angle between the slope ofthe side ribs and the foot end section of the coffin. It will be seen (Fig F7) that the ribs at both the head and foot ends had been tapered to allow for the angles of the sheets.

There are two types of rivet used on this coffin: small-headed rivets have been used simply in the fixing of sheet metal joints; large-headed rivets, in two lengths, have been used to join sheet metal to stiffening ribs. It is clear that the rivets were closed from the outside of the coffin, and this was probably done with a hammer and punch, or a machine press. However, it is much less clear at what point during construction the holes for the rivets were made. Nevertheless, it seems likely that these holes would have been pierced from the inside (the side sheets are distorted outwards at these points) and that they were pierced simultaneously through both the ribs and the side sheets, in order to make sure that the holes were aligned for the proper passage of the rivets.

Design and construction

The most striking feature of the coffin is the inconsistency shown in design elements and quality of construction. Some difficulties must certainly have been posed by the shape of the coffin: the inward sloping of the head end (and probably also the foot end) would have made the creation of correct angles for the side sheets difficult. In addition, the angles which form the four horizontal corners make the forming more complicated.

As Figure F7 shows, in modern terms the head end joint has been inefficiently made, since it would have been much easier to construct the joint with the flush seam on the inside. The interior stiffening rib has had to be bent to accommodate this feature; since it meant more work for the craftsman, it is likely that this feature was included purely for aesthetic reasons. In contrast the foot end exhibits a simple overlap, which shows that the head end was made first and the side sheets finally closed at the foot end.

As shown in Figure F7, the joining technique used to connect the end and side stiffening ribs is also unusual. The end rib has been forged to fit the corner, thus requiring that the side rib also had to be forged. An obvious alternative is shown alongside. It has the additional advantage of making the joint flexible. It seems likely that the technique employed was the result of inexperience or a simple mistake. It should be noted that the radii of the angles formed at the shoulders are different, which gives the coffin a generally lop-sided appearance from the top.

These characteristics indicate that this coffin was not produced during a regular manufacturing process, but was a 'one- off'. As Figure F7 shows, secure coffins are known from this period, but our present knowledge of this section of the funeral industry is so scant that any further conclusions would be rash. Further analytical work on the coffin and on the history of tooling and machinery in this period would undoubtedly be productive.

Appendix G

Entomological and parasitological investigations on samples from Christ Church, Spitalfields, London (a summary)

Andrew K G Jones and J Phipps, Environmental Archaeology Unit, York

The full archive report has been deposited with the Ancient Monuments Laboratory.

Samples of insect remains and fluid and organic sludge collected from lead coffins were submitted for identification and comment. The bulk of the insect remains were puparia of a muscid, probably *Ophyra capensis*, a species which in the larval form is thought to feed on other dipterous larvae. Puparia of coffin flies, *Conicera tibialis*, were found in one sample and a few other taxa were recognized. Fragments of the beetle *Rhizophagus parallelcollis*, a species often associated with human corpses as its larvae feed on wood, were found in two of the samples. No parasite ova were seen in the sludge samples.

The authors of this volume, along with the investigators themselves, had expected more fruitful results from these investigations, and asked Andrew Jones to comment on its implications for future collection of samples from such potentially rich burial sites. His resultant short paper was published in the Newsletter of the Archaeological Formation Processes Study Group (Arch-Form) for February 1988.

Appendix H

Burial catalogue codes (see microfiche M4 for complete catalogue)

Field name	Caption	Code	Description
CN	Context no	0000	Sequential nos
SK	Burial no	0000	Prefix: 2. Based on skeletal sequence; 5000 Numbers are those without human remains; those entries with a skeleton no only belong to other burials
BD	Burial date	0000	See SCRIP for detail
AG	Age	00	In years
EXTA	Coffin plate	0000	Type of plate
EXTADES	Plate design	000	Design typology no
EXTBDES	Lid motif design	000	Design typology no
EXTBDEF	Lid motif design	000	Design typology no
EXTBDEK	Lid motif design	000	Design typology no
EXTBDEB	Lid motif design	000	Design typology no
EXTBDEZ	Lid motif design	000	Design typology no
EXTCDES	Escutcheon design	000	Design typology no
EXTCDEF	Escutcheon design	000	Design typology no
EXTCDEK	Escutcheon design	000	Design typology no
EXTDDES	Grip plate design	000	Design typology no
EXTDDEF	Grip plate design	000	Design typology no
EXTEDES	Grip design	000	Design typology no
EXTEDEF	Grip design	000	Design typology no
AC	Area code	EP	Eastern parochial
		HV	Hebert vault
		LC	Lower central east
		LE	Lower north-east
		LL	Laschallas
		LN	Lower north-west
		LP	Lower portico
		LS	Lower south-west
		LV	Lemaistre
		LW	Lower central west
		NC	North chasm
		NE	Upper north-west
		NP	North parochial
		PK	Peck
		PV	Parochial
		SC	South chasm
		SE	Lower south-east
		SV	Simpson
		UC	Upper central

Field name	Caption	Code	Description
UE	Upper south-east		
		UN	Upper north-west
		US	Upper south-west
UP	UPPN type no	000	Design typology no
LN	Lead type no	01	I.i
		02	I.ii
		03	I.iii
		04	I.iv
		05	II.i
		06	II.ii
		07	III.i
		08	III.ii
		09	III.iii
BP	SK position	01	Supine, extended, hands by sides
		02	As 01; hands over pelvis
		03	As 01; arms crossed
		04	As 01; legs crossed
		05	As 01; arms and legs crossed
		06	Extended, prone
		07	Disturbed in antiquity
		08	No information
CP	Coffin position	01	Horizontal W\E
		02	Horizontal E\W
		03	Horizontal N\S
		04	Horizontal S\N
		05	Horizontal other
		06	Vertical head up
		07	Vertical head down
SC	SK condition	00	Percentage of measurements obtained
MC	Matrix code	A1	1 letter, 1 no
FACTOTAL	Facility type	01	1 shell: wood outer
		02	1 shell: wood inner
		03	2 shell: wood outer, wood inner
		31	2 shell: wood outer, wood inner, lid
		35	1 shell: lead inner
		36	2 shell: wood outer, lead inner
		37	2 shell: lead inner, wood outer
		38	3 shell: wood outer, lead inner, wood inner, lid
		40	Wooden box
		42	Wood outer box, wood inner
		45	Wooden outer, lid
		55	No container
		62	2 shell: trapezoidal wooden outer, wooden inner
		65	2 shell: lead inner, wood inner, lid
		66	Wood outer, lead inner, wood outer, lid
		79	2 shell: lead outer, wood inner
		99	1 shell: ferrous
SEX	Sex	0	M or F or ?
SCRIP	Inscription		From coffin plate
PM	Interval	00	Interval between death and burial in days
STATUS	P\S	0	Primary or secondary

Bibliography

Adams, M & Reeve J, 1987 Excavations at Christ Church, Spitalfields, 1984–6, *Antiquity*, **61**, 232, 247–56

Aitkin, W C, 1866 Coffin furniture manufacture, in *The resources, products and industrial history of Birmingham and the Midland hardware district: a series of reports collected by the local industries committee of the British Association at Birmingham, in 1865*. 704–8

Aughtie, G, 1796 Patent Office, London

Baxby, C, 1985 *Lancet*, 291

Beattie, O, and Geiger, J, 1989, *Frozen in Time: The Fate of the Franklin Expedition* Grafton, London

Binford, L R, 1971 Mortuary practices: their study and their potential, in J A Brown (ed), *Approaches to the social dimensions of mortuary practices*. 6–29

Bishop, H, 1978 Excavations at the church of SS Peter and Paul, Healing, *Lincolnshire Hist Archaeol*, **13**, 81

Boddington, A, 1987 Chaos, disturbance and decay in an Anglo-Saxon cemetery, in A Boddington, A N Garland & R C Janaway (eds), *Death, decay and reconstruction*, 27–42, Manchester: Manchester University Press

Boore, E J, 1985 Excavations at St Augustine the Less, Bristol, 1983–84, *Bristol Avon Archaeol*, **14**, 21–33

Bowis, M, 1985 *Complementary historical report for the excavations of the private and parochial vaults at Christ Church and All Saints, Spitalfields*, unpublished

Brothwell, D, 1987 Decay and disorder in the York Jewbury skeletons, in A Boddington, A N Garland & R C Janaway (eds), *Death, decay and reconstruction*. 22–6 Manchester: Manchester University Press

Butler, L, 1978 St Martin's church, Allerton Mauleverer, *Yorkshire Archaeol J*, **50**, 185

Caldwell, D H, 1976 A group of post-medieval noble burials at Heddington, *Trans E Lothian Antiq Fld Natur Soc*, **15**, 25–37

Cambell, R, 1747 *The London Tradesman*

Carpenter, R, 1746–7 *Purchase ledger*

——, 1761–78 *Sundry accounts*

Chadwick, E, 1843 *A supplementary report on the results of a special inquiry into the practice of interment in towns* House of Lords, London

Chapman R, Kinnes, I & Randsborg, K, 1981 *The archaeology of death* Cambridge: Cambridge University Press

Church, R A & Smith, B M D, 1966 Competition and monopoly in the coffin furniture industry, *Econ Hist Rev*, **19**, 612–41

Clark, S G, and Longhurst, E E, 1961 The Corrosion of Metals by Acid Vapours from Wood *Journal of Applied Chemistry* 11 (1961) 435-443

Colvin, M H, 1978, *A biographical dictionary of British architects*, 1600–1840

Cooper, P, 1974 *Poisoning by drugs and chemicals, plants and animals*, Alchemist publications, London

Crook, J M, 1978 Introduction, in Mortlake, C L (ed J M Crook), *A history of the Gothic Revival in England* [1872]. Leicester: Leicester University Press

Curl, J S, 1972 *The Victorian celebration of death*, David and Charles, Newton Abbott

Deeds, 1713 *Deed, 24th July 1731, Christ Church, Spitalfields*, Lambeth Palace Library

Dethlefson, E & Deetz J, 1966 Deaths' heads, cherubs and willow trees: experimental archaeology in colonial cemeteries, *Amer Antiq*, **31** (4), 502–10

Dezallier-D'Argenville, A J, *Travels et Tours* (Paris *c*1731) London, Victoria and Albert Museum (National Art Library) mss, Press Mark 86 N

Dickens, C, 1853 *Bleak House* (Penguin Books edn, 1971)

Dixon, B, 1985 Doubts that reach from the grave, *New Scientist*, **58**

Edwards, G, 1974 *The Preservation of Textiles in Archaeological Contexts* unpublished diploma dissertation, Institute of Archaeology, London

Evans, W E D, 1963a Adipocere formation in a relatively dry environment, *Medicine, science and the law*, **3**, 145–153

——, 1963b *The Chemistry of Death: American Lecture Series* Charles C. Thomas, Illinois

Garanger, A, 1968 Industrial mechanisation, in M Daumas (ed), *Histoire generale des techniques*, 247–282, J Murray, London

Garland, A N, 1987 A histological study of archaeological bone composition, in A Boddington, A N Garland & R C Janaway (eds), *Death, decay and reconstruction* 109–26 Manchester: Manchester University Press

——, and Janaway, R C, 1989 The Taphonomy of Inhumation Burials, in C A Roberts, F Lee, & J L Bintliff (eds) *Burial Archaeology: Current Research, Methods and Developments* BAR Brit Ser 211 15-37

——, 1988 Histology, taphonomy and human skeletal remains, *Arch-Form*, **2**, 12–16

Gentle, R & Field, R, 1974 *English Domestic Brass*
George, M D, 1925 *London life in the eighteenth century*, London
Gifford, D P, 1981 Taphonomy and paleoecology: A critical review of archaeology's sister disciplines, in M Schiffer, (ed) *Advances in archaeological method and theory*, 465–538
Gittings, C, 1984 *Death, burial and the individual in early modern England* Croom Helm, London
Glover, J M, 1990 The Conservation of Medieval and Later Shrouds from Burials in North West England In: S.A. O'Connor and M. Brooks (eds) *Archaeological Textiles* UKIC Occasional Paper no.10 49-58
Goldstein, L, 1981 One-dimensional archaeology and multi-dimensional people: spatial organisation and mortuary analysis, in R Chapman, I Kinnes & K Randsborg (eds), *The archaeology of death*. 53–69 Cambridge: Cambridge University Press
Henderson, J, 1987 Factors determining the state of preservation of human remains, in A Boddington, A N Garland & R C Janaway (eds), *Death, decay and reconstruction*. 43–54 Manchester: Manchester University Press
Hewer, R H, 1954 The sack-em-up men, *St Bartholemew's Hospital J*, June
Hopkins, C, 1985 Smallpox entombed, *Lancet*, **175**
Jakes, K A and Sibley, P, 1983, Survival of Cellulosic Fibres in the Archaeological Context *Science and Archaeology* 25, 31-38
Janaway, R C, 1989a Corrosion Preserved Textile Evidence: Mechanism, Bias and Interpretation In: R.C. Janaway and B. Scott (eds) *Evidence Preserved in Corrosion Products: New Fields in Artefact Studies* UKIC Occasional Paper No. 8 21-29
——, 1989b The Preservation of Organic Material in Association with Metal Artefacts Deposited in Inhumation Graves In: A. Boddington, A.N. Garland and R.C. Janaway (eds) *Death, Decay and Reconstruction* Manchester University Press 127-148
Johnstone's *London Commercial Directory*, 1834
Justa, P, 1989, The Application of Radiation Methods in the Preservation of Objects of Cultural Heritage Proceedings Technical Congress '89: *Conservation of Historical Monuments and their Role in Modern Society* 90-100
Kranz, P M, 1974 Computer simulation of fossil assemblage formation under conditions of anastrophic burial *J Paleontol*, **48**, 800–08
Kristiansen, K (ed), 1987 *Archaeological formation processes. The representivity of archaeological remains from Danish prehistory* Nationalmuseet, Copenhagen
Last and Jessamine, 1985 Survival of smallpox virus, *Lancet*, **291**
Lewis, W, 1850 *Report of a general scheme for extramural sepulture*. London: HMSO
——, 1851 On the chemical and general effects of the practice of interment in vaults and catacombs, *The Lancet*, **2**, 125–6
Litten, J W S, 1985 Post-medieval vaults: their construction and contents, in *Bull CBA Churches Committee*, **23**, 9–17
——, 1991 *The English way of death: the common funeral since 1450*, Robert Hale, London
Lowery, P R, Savage, R D A & Wilkins, R L, 1971 Scriber, graver, scorper, tracer: notes on experiments in bronzeworking technique, *Proc Prehist Soc,* **37** (1), 167–82
MacPhail, A, 1914 Body snatchers and after. Part 1. A plea for anatomy, *St Bartholemew's Hospital J*, April
Mant, A K, 1987 Knowledge acquired from post war exhumations, in A Boddington, A N Garland & R C Janaway (eds), *Death, decay and reconstruction*. 65–80 Manchester: Manchester University Press
Meggitt, M J, 1965 *The lineage system of the Mae-Enga of New Guinea* Oliver and Boyd, Edinburgh and London
MOL (Museum of London), 1980 *Site manual Part 1: The written record*, London
Morley, J, 1971 *Death, Heaven and the Victorians*, Studio Vista, London
O'Shea, J M, 1981 Social configuration and the archaeological study of mortuary practice; a case study, in R Chapman *et al*, *The archaeology of death*
——, 1984 *Mortuary variability. An archaeological investigation*, Academic Press, London
Pattern books, 1783 and 1826 *Coffin furniture pattern books*, Tuesby & Cooper, 1783, and Anon, 1826, Victoria and Albert Museum collection, M63e and M63f
Phillips, A D, 1976 Excavation techniques in church archaeology, *CBA Res Rep*, **13** (ed P V Addyman & R Morris), 54–63
Plume, Sable (pseudonym), 1910 *Coffins and coffin making*
Polson, C.J, Gee, D J & Knight, B, 1985 *The Essentials of Forensic Medicine* 4th edn Pergamon Press, Oxford
Port, M H (ed), 1986 *The Commissions for Building Fifty New Churches. The minute books 1711–27, a calendar* London Record Society, London
Poulson, C J & Marshall, T K, 1975 *The Disposal of the dead* English University Press, London
Puckle, B S, 1926 *Funeral customs, their origin and development* T Werner Laurie, London
Pugin, A W N, 1844, *Glossary of ecclesiastical ornament and costume* **77**, London
Rahtz, P A, 1981 Artefacts of Christian death, in S C Humphreys & H King (eds), *Mortality and immortality*. 117–136 Academic Press: London
Ralph, B, 1734, *A Critical Review of the Publick Building, statues and ornaments in and about London and Westminster*, London

Raistrick, A, 1972 *Industrial archaeology* Eyre Methuen, London
Razzell, P, 1976 Smallpox extinction – a note of caution, *New Scientist*, July, 35
Redknapp, M, 1984 Little Ilford, St Mary the Virgin, *London Archaeologist, 1984*, 31–7
Reeve, J, 1984 *Christ Church & All Saints, Spitalfields, London E1. Procedure for the recording, excavation and sampling of the vaults (Nov 1984)*. Unpublished
——, & Adams M, 1986 *Christ Church, Spitalfields Archaeological Project, post excavation research design sampling of the vaults*
Rodwell, W, 1981 *The archaeology of the English church*, Batsford, London
Rostoker, W, and Dvorak, J R, 1977 *Interpretation of metalographic structures*, London Academic Press, New York
Rowe, D J, 1983 *Lead manufacturing in Britain: a history* Croom Helm, London
Saxe, A A, 1970 *Social dimensions of mortuary practice*. PhD dissertation, University of Michigan
Schiffer, M B, 1976 *Behavioural archaeology* London Academic Press, New York
——, 1987 *Formation processes of the archaeological record*, University of New Mexico Press, Albuquerque
Sheppard, F W H (ed), 1957 *Survey of London, vol XXVII: Spitalfields and Mile End New Town*
Shipman, P, 1981 *Life history of a fossil: an introduction to taphonomy and palaeoecology*, Harvard University Press, Cambridge
Shoesmith, R, 1980 LLangar, *Archaeol Cambrensis*, **129**, 64–132
Sibley, P, & Jakes, K A, 1984 Survival of Protein Fibres in the Archaeological Context *Science and Archaeology* 26 17-27
Smith, A, 1981–2 A history of tinplate, *Product Finishing*, **34** (12), 38–9; **35** (1), 43–4; **35** (2), 32–3; **35** (3), 62–3; **35** (4), 19–39; **35** (5), 45–6; **35** (6), 51–3; **35** (7), 48–9
Table of fees, 1729 *Table of fees of Christ Church, Spitalfields*, at Greater London Record Office
Tapp, E, & O'Sullivan, D, 1982 St Bee's Man: the Autopsy Proceedings of the Paleopathology Association, 4th European Meeting, Middelberg 178-182
Taylor, J, 1816 *The danger of premature interment proved from many remarkable instances of people who have recovered after being laid out for dead and of others entombed alive for want of being properly examined prior to interment* London
Thomas, C, 1971 *The early Christian archaeology of northern Britain* Oxford University Press for University of Glasgow, London
Thompson, N P & Ross H, 1973 Excavations at the Saxon Church, Alton Barnes, *Wiltshire Archaeol Mag*, **68**, 75–6
Tuesby & Cooper, 1783 *Coffin furniture pattern book*, Victoria and Albert collection, M63e
Turner, J, 1838 *Burial fees of the principal churches, chapels & new burial grounds in London and its environs*, London
Tylecote, R F, 1976 *A history of metallurgy* Metals Society, London
—, 1986 *The prehistory of metallurgy in the British Isles* Institute of metals, London
Urban, J, & Justa, P, 1986 Conservation by Gamma Radiation in the Museum of Central Bohemia in Roztoky *Museum* 38, no3 167-67
Waldron T, 1987 The relative survival of the human skeleton: implications for palaeopathology, in A Boddington, A N Garland & R C Janaway (eds), *Death, decay and reconstruction*. 55–64 Manchester University Press, Manchester
Watts, L & Rahtz, P A, 1985 *Mary-le-Port, Bristol, Excavations 1962–63* City of Bristol Museum and Art Gallery, Bristol
Wildesen, L, 1982 The study of impacts on archaeological sites *Advances in archaeological method and theory*, 5, 51–96
Willats, R M, 1987 Iron graveslabs: a sideline of the early iron industry, *Sussex Archaeol Collect*, **125**, 99–113
Wood, W R & Johnson, D L, 1978 A survey of disturbance processes in archaeological site formation,in *Advances in archaeological method and theory*, ed M Schiffer, **1–4**, 315–83
Worsaae, J J A, 1843, *Denmarks Oldtid oplyst ved Oldsager og Grauoie Selskabet* for Trykkefrihedens nette, Brug, Copenhagen
——, 1849 *The primeval antiquities of Denmark*, John Henry Parker, London
Zuckerman, A J, 1984 Palaeontology of smallpox, *The Lancet*, **22/29,** 1454

Index

compiled by Lyn Greenwood

Material evidence has been indexed under the name of the individual where possible. Italic page numbers indicate figures, bold page numbers indicate tables.